Energy Systems in Electrical Engineering

Series Editor

Muhammad H. Rashid, Florida Polytechnic University, Lakeland, USA

Energy Systems in Electrical Engineering is a unique series that aims to capture advances in electrical energy technology as well as advances electronic devices and systems used to control and capture other sources of energy. Electric power generated from alternate energy sources is getting increasing attention and supports for new initiatives and developments in order to meet the increased energy demands around the world. The availability of computer–based advanced control techniques along with the advancement in the high-power processing capabilities is opening new doors of opportunity for the development, applications and management of energy and electric power. This series aims to serve as a conduit for dissemination of knowledge based on advances in theory, techniques, and applications in electric energy systems. The Series accepts research monographs, introductory and advanced textbooks, professional books, reference works, and select conference proceedings. Areas of interest include, electrical and electronic aspects, applications, and needs of the following key areas:

- Biomass and Wastes Energy
- Carbon Management
- Costs and Marketing
- Diagnostics and Protections
- Distributed Energy Systems
- Distribution System Control and Communication
- Electric Vehicles and Tractions Applications
- Electromechanical Energy Conversion
- Energy Conversion Systems
- Energy Costs and Monitoring
- Energy Economics
- Energy Efficiency
- Energy and Environment
- Energy Management, and Monitoring
- Energy Policy
- Energy Security
- Energy Storage and Transportation
- Energy Sustainability
- Fuel Cells
- Geothermal Energy
- Hydrogen, Methanol and Ethanol Energy
- Hydropower and Technology
- Intelligent Control of Power and Energy Systems
- Nuclear Energy and Technology
- Ocean Energy
- Power and Energy Conversions and Processing
- Power Electronics and Power Systems
- Renewable Energy Technologies
- Simulation and Modeling for Energy Systems
- Superconducting for Energy Applications
- Tidal Energy
- Transport Energy

The books of this series are reviewed in a single blind peer review process by the main editor and/or the advisory board.

Iman Rahimi · Amir H. Gandomi ·
Miadreza Shafie-khah · Abdollah Ahmadi
Editors

Sustainable Supply Chain of Renewable Energy Networks

Springer

Editors
Iman Rahimi
Data Science Institute
University of Technology Sydney
Sydney, NSW, Australia

Amir H. Gandomi
Data Science Institute
University of Technology Sydney
Sydney, NSW, Australia

Miadreza Shafie-khah
Research and Innovation Division
Nowocert, Dublin, Ireland

Abdollah Ahmadi
Ampol
Alexandria, NSW, Australia

ISSN 2199-8582 ISSN 2199-8590 (electronic)
Energy Systems in Electrical Engineering
ISBN 978-981-96-4939-6 ISBN 978-981-96-4940-2 (eBook)
https://doi.org/10.1007/978-981-96-4940-2

This Springer imprint is published by the registered company Springer Nature Singapore Pte Ltd.
The registered company address is: 152 Beach Road, #21-01/04 Gateway East, Singapore 189721, Singapore

This book
is dedicated to

My Parents, who have always supported me throughout my life.

—Dr. Iman Rahimi

My parents, who are constant sources of support through all of my life's challenges.

—Dr. Miadreza Shafie-khah

My parents, who never stopped believing in me.

—Dr. Abdollah Ahmadi

Preface

The growing urgency of climate change and its impact on humanity has compelled researchers, practitioners, and policymakers to explore innovative solutions for sustainable energy systems. This edited book, *Sustainable Supply Chain of Renewable Energy Networks*, brings together critical advancements in renewable energy systems and their supply chains. It provides a multidisciplinary platform for understanding the challenges and opportunities within the global energy transition, emphasizing optimization, innovation, and sustainability.

The book explores a wide array of topics, including foundational concepts in renewable energy systems, energy management strategies, cutting-edge optimization techniques, and advanced technologies for energy planning. Each chapter contributes to an overarching framework aimed at aligning energy systems with Sustainable Development Goals (SDGs), promoting social justice, and achieving global decarbonization objectives.

This volume comprises 10 chapters, each authored by leading experts and practitioners from across the globe. It begins with an insightful overview of the role of sustainable supply chains in renewable energy systems and transitions seamlessly into specialized topics such as energy management, optimization techniques, and integrated systems for power and natural gas. Later chapters delve into the applications of novel technologies, including blockchain, artificial intelligence, and decision support systems, for enhancing energy efficiency and resilience. The book also addresses practical challenges, such as supply chain disruptions, and introduces innovative frameworks for modeling tools and optimization approaches.

One of the unique strengths of this book is its global perspective. With contributions from diverse regions, including Europe, Asia, and Africa, the book highlights successful case studies and collaborative initiatives that underscore the importance of international cooperation. Examples of renewable energy projects, such as offshore wind farms, solar networks, and polygeneration platforms, illustrate how technological advancements can address both local and global energy challenges.

This edited volume is intended for a wide audience, including academic researchers, engineers, policymakers, and graduate students. By providing a comprehensive resource on sustainable energy supply chains, the book aims to inspire further research and practical implementations that can lead to a cleaner, more equitable, and sustainable energy future. It serves as a valuable reference for advancing our collective understanding of renewable energy systems and supporting the transition toward a low-carbon global economy.

We express our heartfelt gratitude to all contributors, co-editors, and reviewers for their commitment to this work. Their efforts have ensured the quality and depth of the chapters included in this book. We also thank the academic and industry communities for their unwavering dedication to solving the pressing challenges of energy sustainability. It is our hope that this book will serve as both a foundational resource and an inspiration for future innovations in renewable energy systems and sustainable supply chains.

We look forward to seeing how these insights and frameworks contribute to the realization of a more sustainable and resilient world.

Sydney, Australia — Iman Rahimi, Ph.D.
Sydney, Australia — Amir H. Gandomi, Ph.D.
Nowocert, Ireland — Miadreza Shafie-khah, Ph.D.
Alexandria, Australia — Abdollah Ahmadi, Ph.D.

Acknowledgements

We, the editors, extend our heartfelt gratitude to all contributors, reviewers, and collaborators who made this book, *Sustainable Supply Chain of Renewable Energy Networks*. Special thanks to the chapter authors for their expertise and dedication, as well as the reviewers for their meticulous evaluations and constructive feedback. We deeply appreciate the editorial and project management team at Springer for their professionalism and unwavering support throughout this journey; your guidance and collaboration have been invaluable in shaping this work. Finally, we acknowledge the unwavering encouragement of our families and colleagues, whose support has been a constant source of strength. This book is a testament to the collective effort and shared vision for a sustainable energy future.

Iman Rahimi, Ph.D.
Amir H. Gandomi, Ph.D.
Miadreza Shafie-khah, Ph.D.
Abdollah Ahmadi, Ph.D.

Contents

Editors and Contributors

About the Editors

Iman Rahimi Ph.D. is a data scientist and optimization expert with a robust background in mathematics and industrial engineering. He holds dual Ph.D.s: one in Industrial Engineering from Universiti Putra Malaysia and another in Computer Science from the University of Technology Sydney. His expertise includes machine learning, evolutionary algorithms, and large-scale optimization, with applications in renewable energy, supply chain management, and computational intelligence. Dr. Rahimi has received several prestigious international awards and actively serves as a reviewer for leading journals. His research contributions are widely published in high-impact journals, reflecting his commitment to advancing data-driven decision-making and AI-driven optimization.

Amir H. Gandomi is recognized as one of the world's most cited researchers in the fields of global optimization and big data analytics, particularly for his work using machine learning and evolutionary computations. He is an ARC DECRA fellow at the Faculty of Engineering and Information Technology at the University of Technology Sydney, where he serves as a professor of Data Science. He has received multiple prestigious awards for his research excellence and impact, including the 2024 IEEE TCSC Award for Excellence in Scalable Computing (MCR), the 2023 Achenbach Medal (awarded at Stanford University), and the 2022 Walter L. Huber Prize, which is the highest-level mid-career research award in all areas of civil engineering. Amir has published over 450 journal papers and 14 books. He has been named one of the world's most influential scientific minds and highly cited researchers by Clarivate Analytics for six consecutive years.

Miadreza Shafie-khah is a full professor and the scientific director of the Energy Business eMBA program at the University of Vaasa Executive Education in Vaasa, Finland. He also leads a team at the Research and Innovation Division of Nowocert in Dublin, Ireland. He has served as Editor-in-Chief and associate editor for several

IEEE Transactions and journals. He is the Principal Investigator (PI) or Co-Principal Investigator (Co-PI) of several international and EU projects, and he was ranked as the number one "Rising Star of Science" in Finland for 2022–2024. Prof. Shafie-khah has published over 550 scientific articles and he has received five Best Paper Awards at IEEE Conferences and five highly cited papers recognized by ISI Web of Science.

Abdollah Ahmadi is the energy optimization lead at Ampol Australia. As a renewable energy scholar, he has published more than 85 international journal papers and book chapters. His research focuses on the applications of novel energy storage technologies in renewable energy-integrated electric power systems and virtual power plants (VPP) for large-scale and distribution grids in the National Electricity Market (NEM).

Contributors

Mohammad Reza Aghamohammadi Faculty of Electrical Engineering, Shahid Beheshti University, Tehran, Iran

Abdollah Ahmadi Ampol Australia, Sydney, Australia

Theyab R. Alsenani Department of Electrical Engineering, College of Engineering in Al-Kharj, Prince Sattam Bin Abdulaziz University, Al-Kharj, Saudi Arabia

Hossein Ameli Control and Power Group, Imperial College London, London, UK

Mohammad Taghi Ameli Faculty of Electrical Engineering, Shahid Beheshti University, Tehran, Iran

Mohammad Mehdi Amiri Faculty of Electrical Engineering, Shahid Beheshti University, Tehran, Iran

Ismail Baklouti National School of Engineers of Monastir, University of Monastir, Monastir, Tunisia

Sudip Dey Department of Mechanical Engineering, National Institute of Technology, Silchar, Assam, India

Amir H. Gandomi Faculty of Engineering and Information Technology, University of Technology Sydney, Ultimo, Australia;
University Research and Innovation Center (EKIK), Óbuda University, Budapest, Hungary

Mostaan Khakpoor Department of Electrical and Computer Engineering, University of Michigan-Dearborn, Dearborn, USA

Vamsi Krishna Department of Mechanical Engineering, Sri Sivasubramaniya Nadar (SSN) College of Engineering, Kalavakkam, Tamil Nadu, India

Tommi Lehtonen University of Vaasa, Innovation and Entrepreneurship InnoLab, Vaasa, Finland

Saikat Ranjan Maity Department of Mechanical Engineering, National Institute of Technology, Jamshedpur, Adityapur, Jharkhand, India

Iman Rahimi Faculty of Engineering and Information Technology, University of Technology Sydney, Ultimo, Australia

Saleem ur Rahman Rabat Business School, The International University of Rabat (Université Internationale de Rabat), Rabat, Morocco

Homa Rashidizadeh-Kermani Department of Electrical and Computer Engineering, University of Birjand, Birjand, Iran;
Research and Innovation Division, Nowocert, Dublin, Ireland

Mostafa Shabanian-Poodeh Department of Electrical Engineering, Faculty of Engineering, University of Isfahan, Isfahan, Iran

Miadreza Shafie-khah Research and Innovation Division, Nowocert, Dublin, Ireland

Shah Rukh Shakeel School of Marketing and Communication, and Innovation and Entrepreneurship InnoLab, University of Vaasa, Vaasa, Finland

Pierluigi Siano Department of Management and Innovation Systems, University of Salerno, Salerno, Italy

Rishav Raj Singh Department of Mechanical Engineering, National Institute of Technology, Silchar, Assam, India

Goran Strbac Control and Power Group, Imperial College London, London, UK

Siamak Talathahari Faculty of Engineering and Information Technology, University of Technology Sydney, Ultimo, Australia;
Department of Computer Science, Khazar University, Baku, Azerbaijan

Divya Zindani Department of Mechanical Engineering, Sri Sivasubramaniya Nadar (SSN) College of Engineering, Kalavakkam, Tamil Nadu, India

Chapter 1
Sustainable Supply Chain of Renewable Energy: Foundations, Innovations, and Global Perspectives

Iman Rahimi, Siamak Talathahari, Abdollah Ahmadi, Miadreza Shafie-khah, and Amir H. Gandomi

Abstract A sustainable supply chain plays a critical role in advancing renewable energy systems, which emphasizes innovation, global collaboration, and interdisciplinary research. The focus on this research area has been increased dramatically. Academic contributions are dominated by topics like "Energy Fuels" and "Green Sustainable Science Technology." Prominent establishments such as Imperial College London and Iran University spearhead these initiatives, demonstrating worldwide involvement. The efficiency of resource use, waste reduction, and energy distribution has greatly increased with the integration of cutting-edge technologies like blockchain, smart grids, and artificial intelligence. International cooperation has also been crucial; successful information exchange is demonstrated by European offshore wind farms and African solar projects. Future directions center on connection with the Sustainable Development Goals of the UN, circular economy concepts, and regional adaptations. A robust and equitable future for renewable energy is made possible by sustainable supply chains, which tackle technological, financial, and environmental issues.

Keywords Sustainable · Renewable energy · Supply chain · Innovation

---

I. Rahimi (✉) · S. Talathahari · A. H. Gandomi
Faculty of Engineering and Information Technology, University of Technology Sydney, Ultimo, Australia
e-mail: Iman83@gmail.com

S. Talathahari
Department of Computer Science, Khazar University, Baku, Azerbaijan

A. Ahmadi
Ampol Australia, Sydney, Australia

M. Shafie-khah
Research and Innovation Division, Nowocert, Dublin, Ireland

A. H. Gandomi
University Research and Innovation Center (EKIK), Óbuda University, Budapest, Hungary

I. Rahimi et al. (eds.), *Sustainable Supply Chain of Renewable Energy Networks*, Energy Systems in Electrical Engineering, https://doi.org/10.1007/978-981-96-4940-2_1

1.1 Introduction to Sustainable Supply Chains

Renewable energy is more important than ever in a world that aims for sustainable growth. As the effects of climate change become more apparent, there is an increasing demand for clean energy sources including solar, wind, marine, and biomass on a worldwide scale. However, the transition from traditional sources to renewable energy is not without its difficulties, especially when it comes to supply chain management. For renewable energy to be produced responsibly and distributed to customers in a way that maximizes efficiency, reduces waste, and supports long-term environmental goals, sustainable supply chains are essential (Hassan et al., 2024; Shrestha et al., 2022).

Because of strict carbon regulations, growing consumer awareness, and environmental preservation, carbon reduction has emerged as one of the most crucial goals in supply chain (SC) design (Abbasi & Choukolaei, 2023). Installation of renewable energy facilities across countries has been studied in the literature (Holloway et al., 2023; Nematian & Rahimi, 2022; Rahimi et al., 2023). However, in renewable energy, a sustainable supply chain encompasses more than just the generation of clean energy. From obtaining resources to producing energy and delivering it to the final consumer, it covers every phase. These procedures are optimized in a sustainable supply chain by taking ecological and economic considerations into account. Supply networks need to adjust to the demands of a constantly shifting environment while also satisfying the demands of a world population that is expanding. A truly renewable energy future depends on supply chain optimization since clean energy is only as sustainable as the means by which it is distributed (Fontes et al., 2018).

Over time, the idea of sustainability itself has changed. Sustainability in the context of renewable energy is defined as lowering carbon emissions, minimizing adverse environmental effects, fostering economic stability, and enhancing the welfare of society (Seuring & Müller, 2008). In contrast to conventional supply chains, which frequently place a higher priority on speed and cost, a sustainable renewable energy supply chain prioritizes durability, resilience, and a small environmental impact (Kim et al., 2014). It's about striking a balance between ensuring that everyone has dependable access to energy and preserving our environment.

Although the renewable energy industry has experienced rapid expansion in recent years, the path has not been easy. Due to regional resources, laws, and economic considerations, different parts of the world have taken different tacks when it comes to renewable energy. While some nations have emerged as pioneers in wind and solar energy, others still face major obstacles, like a lack of funding or infrastructure to make the switch from fossil fuels (Kabel & Bassim, 2020; Mirza et al., 2009). For example, with a heavy emphasis on offshore wind farms and smart grid technology, Europe is leading the way in the development of integrated renewable energy networks (Capizzi et al., 2019; Pacesila et al., 2016). On the other hand, despite infrastructural issues still being a major barrier, several African countries are investigating solar energy as a way to address both the need for economic growth and energy constraints (Adeoye & Spataru, 2020; Francis et al., 2022). China is at the forefront of producing renewable energy components, and Asia has been using its technological breakthroughs to progress massive solar and marine energy projects (Zhang et al., 2017).

Supply chains for renewable energy nevertheless confront challenges despite these initiatives, which need creativity, teamwork, and strategic management. Current supply chains must change in the areas of energy storage, effective distribution, and loss reduction. To successfully handle these issues, supply chains must be optimized to satisfy the particular requirements of renewable energy sources.

This chapter provides a foundation for a sustainable supply chain in renewable energy. Section 1.2 illustrates an overview of the sustainable supply chain of renewable energy. Section 1.3 provides the role of innovation in renewable energy supply chain management. Section 1.4 presents global perspectives and collaborative efforts. Section 1.5 shows the conclusion and future directions.

1.2 An Overview of the Sustainable Supply Chain of Renewable Energy

A key foundation for moving toward a low-carbon economy is the sustainable supply chain of renewable energy, which makes sure that the production, distribution, and storage of renewable energy resources are economical and ecologically responsible. From locating raw materials and producing components to moving, setting up, and maintaining energy infrastructure including solar panels, wind turbines, and biomass plants, it covers every stage of the lifecycle of renewable energy systems. Through the integration of cutting-edge technologies such as blockchain, digital twins, and artificial intelligence, sustainable supply chains optimize energy distribution networks, minimize waste, and improve resource utilization. Additionally, these supply chains support international cooperation and innovation while tackling the issues of decarbonization, economic viability, and energy equity in line with global sustainability goals. In addition to reducing the environmental impact of energy systems, this all-encompassing strategy increases resilience to geopolitical upheavals and climate change, paving the way for a time when everyone can access sustainable clean energy. The field of sustainability is in the center of concentration of many researchers across the world. Table 1.1 presents the ten most cited documents on the sustainable supply chain of renewable energy, illustrating a wide range of influential contributions published in respected journals such as Renewable Energy, Advanced Energy Materials, Journal of Cleaner Production, Sustainable Development, Energy Policy, Journal of Power Sources, Environmental Science and Pollution Research, and the International Journal of Life Cycle Assessment. These studies encompass varied themes, including the selection of renewable energy sources using analytic hierarchy processes for sustainable electricity generation systems (Ahmad & Tahar, 2014), exploring supply chain and logistics challenges in bio-energy production (Gold & Seuring, 2011), developing sodium-ion batteries that facilitate large-scale energy storage for grid applications (Hirsh et al., 2020), and examining emerging green ideologies and their role in sustainable development in Asian economies (Khan et al., 2019).

Figure 1.1 illustrates the publication trends in renewable energy supply chain sustainability from 1999 to 2024, highlighting a significant rise in academic interest over the years. The publication patterns in renewable energy supply chain sustainability from 1999 to 2024 are depicted in the bar chart, which shows a notable increase in scholarly interest over time. Prior to 2015, there was little to no early research effort, with less than ten publications each year. A discernible rise started in 2016, indicating an increased interest in the field, and this trend picked up speed starting in 2020. Publications reached 144, 151, and 221 in 2022, 2023, and 2024, respectively, marking the most significant growth.

Figure 1.2 visualizes the distribution of record counts across different Web of Science categories related to renewable energy research. According identified categories in Fig. 1.2, the greatest number, 433, "Energy Fuels" is in first place, followed by "Green Sustainable Science Technology" (356), and "Environmental Sciences" (302). "Engineering Environmental," "Environmental Studies," and "Engineering Chemical" are some noteworthy categories that highlight the interdisciplinary character of research on renewable energy. This graphic demonstrates the importance of engineering and environmental studies to the field while emphasizing the predominance of energy-focused and sustainability-related categories. The graph highlights the variety and scope of research on renewable energy.

1.3 Interdisciplinary Focus in Renewable Energy Research

Figure 1.3, word cloud visualization, showcases the distribution of Web of Science categories related to renewable energy research based on their record counts. The graphic is dominated by prominent categories like "Energy Fuels," "Green Sustainable," and "Environmental Sciences" because of their important contributions, which highlight how crucial they are to solving the problems associated with renewable energy. These areas are at the forefront of innovation and research, propelling improvements in environmental sustainability, energy system optimization, and energy generation. Sustainability and environmental science are essential elements of study on renewable energy since the size of the category names strongly relates to the number of records in each category.

The multidisciplinary aspect of this profession is shown by smaller but no less significant categories such as "Engineering Environmental," "Environmental Studies," and "Engineering Chemical," which combine chemical processes and engineering concepts to improve energy systems. Furthermore, sections like "Management," "Economics," and "Thermodynamics" draw attention to the variety of research facets, such as the technical effectiveness and socioeconomic aspects of renewable

Table 1.1 Ten most-cited documents on the sustainable supply chain of renewable energy

R	TC	Journal	Title	Author(s)	Year
1	678	Renewable Energy	Selection of renewable energy sources for sustainable development of electricity generation system using analytic hierarchy process: a case of Malaysia (Ahmad & Tahar, 2014)	Ahmad and Taher	(2014)
2	559	Advanced Energy Materials	Sodium-ion batteries paving the way for grid energy storage (Hirsh et al., 2020)	Hirsh et al.	(2020)
3	554	Journal of Cleaner Production	Supply chain and logistics issues of bio-energy production (Gold & Seuring, 2011)	Gold and Seuring	(2011)
4	458	Sustainable Development	A green ideology in Asian emerging economies: from environmental policy and sustainable development (Khan et al., 2019)	Khan et al.	(2019)
5	455	Sustainable Development	Measuring the impact of renewable energy, public health expenditure, logistics, and environmental performance on sustainable economic growth (Khan et al., 2020)	Khan et al.	(2020)
6	438	Energy Policy	Impacts of COVID-19 pandemic on the global energy system and the shift progress to renewable energy: opportunities, challenges, and policy implications	Hoang et al.	(2021)
7	395	Journal of Power Sources	Stochastic control of smart home energy management with plug-in electric vehicle battery energy storage and photovoltaic array (Wu et al., 2016)	Wu et al.	(2016)
8	393	The International Journal of Life Cycle Assessment	LCA of a biorefinery concept producing bioethanol, bio-energy, and chemicals from switchgrass (Cherubini & Jungmeier, 2010)	Cherubini and Jungmeier	(2010)
9	337	Sustainable Development	Investigate the role of technology innovation and renewable energy in reducing transport sector CO_2 emission in China: a path toward sustainable development (Godil et al., 2021)	Godil et al.	(2021)
10	208	Environmental Science and Pollution Research	Environmental logistics performance indicators affecting per capita income and sectoral growth: evidence from a panel of selected global ranked logistics countries Mastrocinque (Khan et al., 2017)	Khan et al.	(2017)

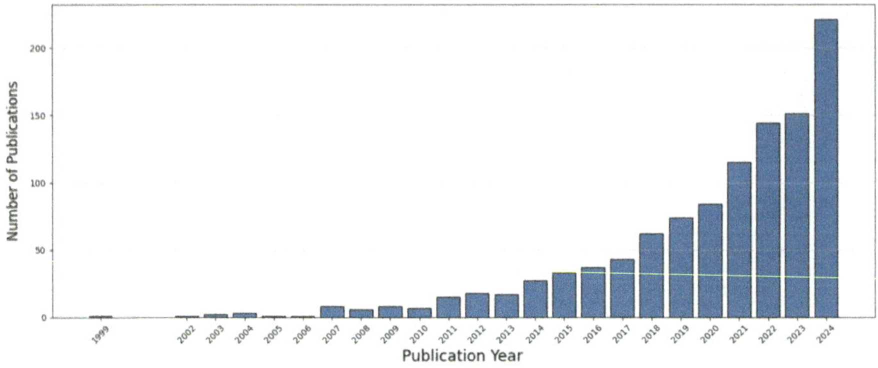

Fig. 1.1 Number of publications on renewable energy supply chain sustainability over the years

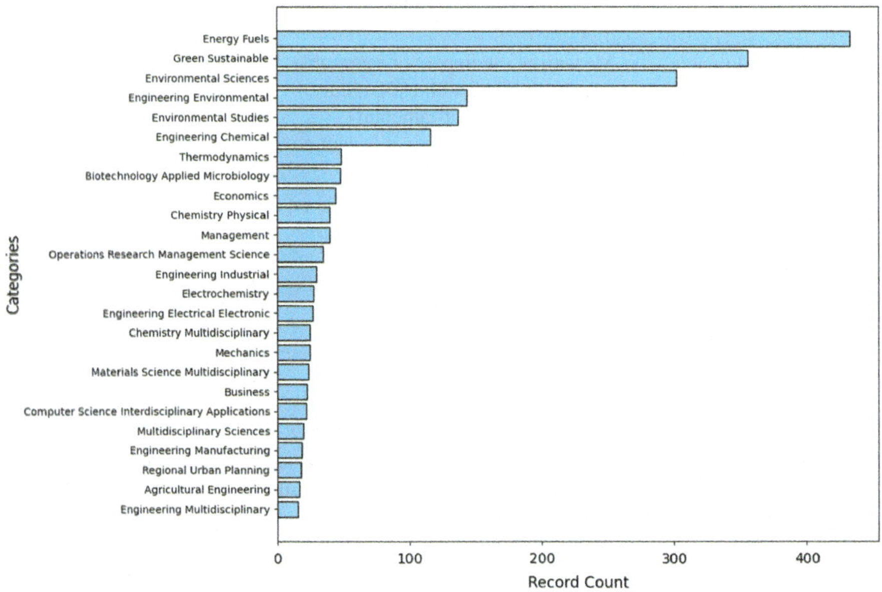

Fig. 1.2 Record counts for renewable energy

energy systems. The range of scholarly interest in renewable energy is well represented by this graphic, which also highlights the need for multidisciplinary and cooperative efforts to address global sustainability issues.

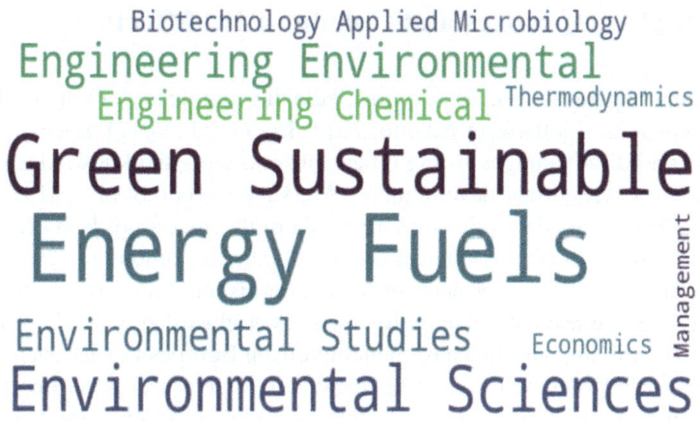

Fig. 1.3 Word cloud of web of science categories

1.4 The Role of Innovation in Renewable Energy Supply Chain Management

Innovation is at the heart of transforming renewable energy supply chains (González et al., 2012; Wang et al., 2024). New opportunities to improve the sustainability and efficiency of energy networks have been made possible by technological developments and optimization strategies (Banos et al., 2011; Potrč et al., 2021). For instance, decision support systems (DSSs) are essential to enhancing the distribution and storage of energy. These systems utilize data analytics, artificial intelligence, and machine learning to enable informed decision-making, allowing for a more flexible and adaptive approach to managing energy supply chains (Beriro et al., 2022; Liu & Du, 2020).

A few examples of developments influencing the future of supply chain management for renewable energy include the usage of smart grids, blockchain for transparency, and sophisticated forecasting models (Hossain et al., 2016; Nia et al., 2021). Blockchain ensures supply chain transparency, lowering fraud and inefficiencies, while smart grids offer real-time data on energy flow, facilitating improved coordination between producers and consumers. Managing intricate renewable energy networks, which are inherently unpredictable and variable, requires the use of such tools (Almutairi et al., 2023; Kushch & Castrillo, 2017).

Material science is another facet of innovation. Examples of how technological developments can directly enhance supply chain sustainability include the creation of more effective solar panels, recyclable wind turbine blades, and energy storage devices. These developments lessen the total environmental impact of renewable energy infrastructure while simultaneously increasing the efficiency of energy production (Cucchiella & D'Adamo, 2013; Mertens et al., 2024).

1.5 Global Perspectives and Collaborative Efforts

International cooperation is necessary to handle the complexity of renewable energy supply chains (Cucchiella & D'Adamo, 2013; Irani et al., 2017). Each country has advantages and disadvantages facing renewable energies, and knowledge exchange can accelerate international development (Aleixandre-Tudó et al., 2019). Stronger and more effective supply chains may result from the sharing of best practices and knowledge from various geographical areas (Aleixandre-Tudó et al., 2019).

For example, cooperative projects in Africa, for instance, have shown how renewable energy may be used to solve regional issues (Belward et al., 2011; Chineke & Ezike, 2010). African countries have benefited from European experience in renewable energy infrastructure by creating localized solar power solutions that offer economic opportunities and electricity access. Other examples are North American technology contributions, which have helped South American biomass energy programs by facilitating more effective processing and delivery (Barbosa et al., 2017; Etcheverry, 2011).

In addition to advancing technology, international cooperation could accelerate coordinating renewable energy projects with global objectives like the Sustainable Development Goals (SDGs) of the UN (Marco-Lajara et al., 2023; McCollum et al., 2018). Moreover, optimizing renewable energy supply chains has a direct influence on several SDGs, including lowering carbon emissions, encouraging responsible consumption, and guaranteeing access to clean energy (Anantha Subramanya Iyer et al., 2024; Emenike & Falcone, 2020; Hiloidhari et al., 2023). Figure 1.4 illustrates the distribution of research records across different countries and regions in the field of renewable energy. China and the USA are the countries with the largest contributions, with 12.1 and 11.8% of the total records, respectively. India (9.3%), England (9.0%), and Italy (8.6%) are other noteworthy donors, demonstrating their strong emphasis on research in renewable energy. While Australia, Malaysia, and the Netherlands have moderate contributions ranging between 3.4 and 5.0%, countries like Germany and Iran each provide 6.2%. With participation from nations in Asia, Europe, the Americas, and Oceania, the contributors' diversity highlights the interest in renewable energy research on a worldwide scale. This graphic emphasizes how cooperative and global efforts are in tackling energy sustainability issues.

1.6 Key Research Areas in Renewable Energy

Figure 1.5 highlights the distribution of research records across the top research areas contributing to renewable energy studies. Energy fuels are in the first place, demonstrating its crucial contribution to the development of renewable energy, with 331, 294 and 284 records, respectively, and Science Technology and Environmental Sciences Ecology come in second and third, highlighting the field's interdisciplinary nature. Two hundred and seventy-six records are related to engineering, with a focus

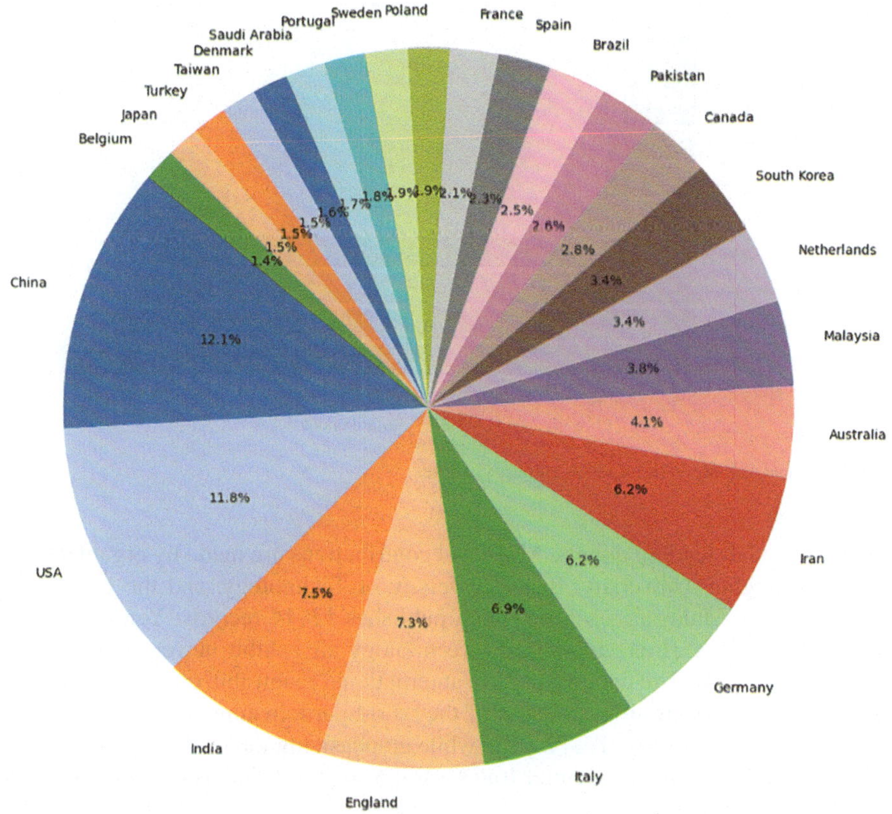

Fig. 1.4 Record distribution by countries/regions

on how technological solutions are incorporated into renewable energy systems. The wide range of research is demonstrated by other fields such as business economics, biotechnology, and computer science, which go beyond conventional energy-focused fields to encompass advances in computation and socioeconomics. This diagram makes it evident how several disciplines work together to tackle the complicated issues surrounding renewable energy.

1.7 Leading Academic Affiliations in Renewable Energy Research

Figure 1.6 provides the top ten affiliations that have the most research productivity in the field. Iran University is at the top, demonstrating its leadership in the area, with more than 22 publications. With record counts ranging from 16 to 20, Imperial College London, Helmholtz Association, and Indian Institute of Technology are

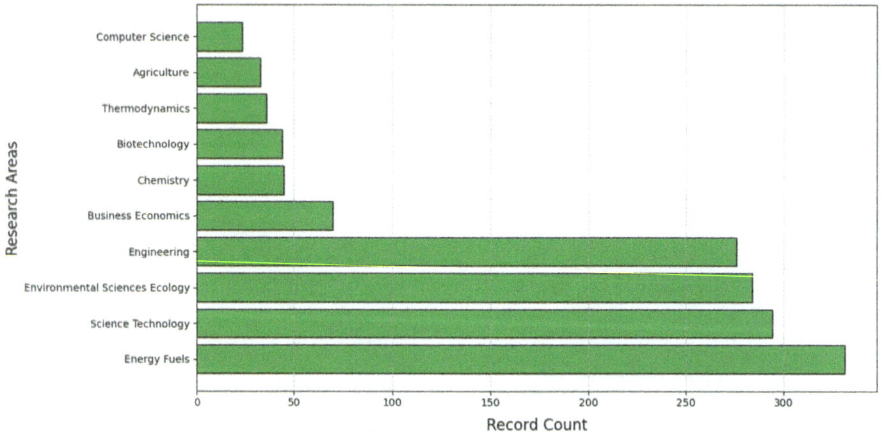

Fig. 1.5 Research areas in renewable energy studies

among the other leading donors. Mid-level contributions are made by organizations such as Tsinghua University, Brno University of Technology, and the University of Nottingham Malaysia, each of which publishes 12–14 records. Yonsei University, RMIT, and Karlruhe Institute of Technology all exhibit newer contributions with fewer records. The data shows how international renewable energy research is, with institutions from different parts of the world participating. The graphic highlights a variety of cooperative projects while emphasizing the leadership of specific colleges. This distribution demonstrates varying research intensities among institutions, emphasizing both regional leaders and emerging players in the renewable energy domain.

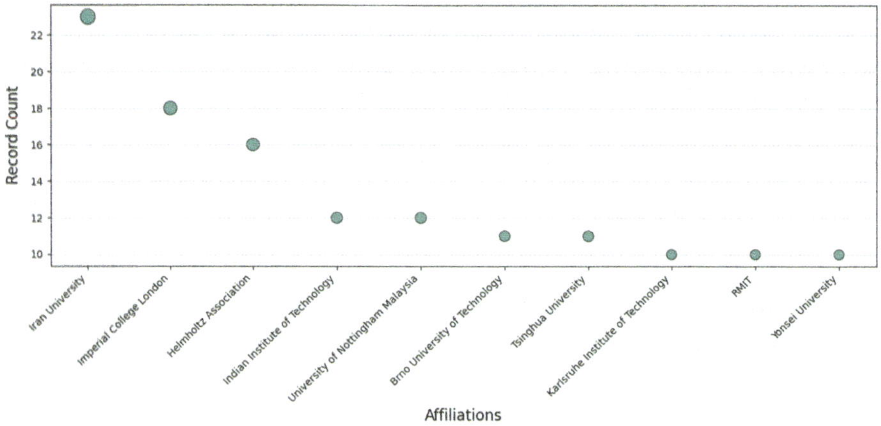

Fig. 1.6 Top affiliations in renewable energy research

1.8 Aligning with Sustainable Development Goals

The United Nations' definition of sustainability and the drive for renewable energy are closely related (Colombo et al., 2013; Dincer, 2000). A crucial element of matching renewable energy projects with these objectives is supply chain optimization. In order to build sustainable communities and advance global well-being, clean energy generation, lower carbon footprints, and fair energy distribution are essential (Banos et al., 2011; Iqbal et al., 2014).

For example, SDG 7 aims to ensure access to affordable, reliable, and sustainable energy for all (Gebara & Laurent, 2023). Producing renewable energy is simply one aspect of achieving this aim; another is ensuring that energy is delivered fairly and efficiently. To ensure that all communities, even those in remote or impoverished areas, benefit from renewable energy, sustainable supply chains are essential to this endeavor. They do this by facilitating the efficient flow of energy from the point of production to the point of consumption.

Supply chain optimization also supports SDG 13, which calls for urgent action to combat climate change. Renewable energy supply chains can drastically reduce the carbon emissions related to energy distribution by cutting waste, increasing efficiency, and utilizing cutting-edge technologies. This makes the supply chain a crucial component of the solution and unites the renewable energy industry with the larger battle against climate change (Li et al., 2019; Raman et al., 2023).

1.9 Conclusion and Future Directions

The presented chapter emphasizes how crucial sustainable supply chains are to hasten the world's switch to renewable energy. Sustainable supply chains facilitate the effective integration of renewable energy systems from production to consumption by tackling the issues of resource optimization, waste reduction, and equitable energy distribution. Supply chains are growing more robust and flexible thanks to cutting-edge techniques like blockchain, artificial intelligence, and sophisticated decision support systems, guaranteeing the smooth flow of renewable energy even in the face of geopolitical and environmental upheavals. Additionally, the multidisciplinary contributions from disciplines like environmental sciences, engineering, and business emphasize the teamwork required to achieve global sustainability goals like lowering carbon emissions, enhancing energy equity, and promoting economic stability.

The chapter highlights the necessity of ongoing innovation and international cooperation to overcome the remaining difficulties in renewable energy supply chains. The future of renewable energy depends on flexible and integrated supply chain systems, from utilizing smart grids and next-generation materials to investigating region-specific solutions. With their distinct resources and experience, nations must collaborate to exchange ideas, create best practices, and match their plans with the Sustainable Development Goals of the UN. A cleaner, more resilient energy future that benefits everyone may be achieved by cultivating such collaborations and placing a high priority on sustainability.

In future work, it is recommended to utilize cutting-edge technology like artificial intelligence (AI), blockchain, and smart grids. Sustainability will be advanced by region-specific plans and circular economy ideas, such as recycling parts. Building resilient and egalitarian energy systems will need stepping up international cooperation and coordinating initiatives with the Sustainable Development Goals of the UN.

References

Abbasi, S., & Choukolaei, H. A. (2023). A systematic review of green supply chain network design literature focusing on carbon policy. *Decision Analytics Journal, 6*, 100189.

Adeoye, O., & Spataru, C. (2020). Quantifying the integration of renewable energy sources in West Africa's interconnected electricity network. *Renewable and Sustainable Energy Reviews, 120*, 109647.

Ahmad, S., & Tahar, R. M. (2014). Selection of renewable energy sources for sustainable development of electricity generation system using analytic hierarchy process: A case of Malaysia. *Renewable Energy, 63*, 458–466.

Aleixandre-Tudó, J. L., Castelló-Cogollos, L., Aleixandre, J. L., & Aleixandre-Benavent, R. (2019). Renewable energies: Worldwide trends in research, funding and international collaboration. *Renewable Energy, 139*, 268–278.

Almutairi, K., Hosseini Dehshiri, S. J., Hosseini Dehshiri, S. S., Hoa, A. X., Arockia Dhanraj, J., Mostafaeipour, A., Issakhov, A., & Techato, K. (2023). Blockchain technology application challenges in renewable energy supply chain management. *Environmental Science and Pollution Research, 30*(28), 72041–72058.

Anantha Subramanya Iyer, K. N., Desai, K., & Kautish, S. (2024). Optimizing green power and green energy through digital technologies. In *Digital technologies to implement the UN sustainable development goals* (pp. 275–304). Springer.

Banos, R., Manzano-Agugliaro, F., Montoya, F. G., Gil, C., Alcayde, A., & Gómez, J. (2011). Optimization methods applied to renewable and sustainable energy: A review. *Renewable and Sustainable Energy Reviews, 15*(4), 1753–1766.

Barbosa, L. de S. N. S., Bogdanov, D., Vainikka, P., & Breyer, C. (2017). Hydro, wind and solar power as a base for a 100% renewable energy supply for South and Central America. *PLoS One, 12*(3), e0173820.

Belward, A., Bisselink, B., Bódis, K., Brink, A., Dallemand, J. F., De Roo, A., Huld, T., Kayitakire, F., Mayaux, P., Moner-Girona, M., et al. (2011). Renewable energies in Africa. *Rapport Technique, European Commission-Joint Research Center, Ispra, Italy, 182*, 183.

Beriro, D., Nathanail, J., Salazar, J., Kingdon, A., Marchant, A., Richardson, S., Gillet, A., Rautenberg, S., Hammond, E., Beardmore, J., et al. (2022). A decision support system to assess the feasibility of onshore renewable energy infrastructure. *Renewable and Sustainable Energy Reviews, 168*, 112771.

Capizzi, F., Das, A., Dauwe, T., Moorkens, I., Juhana, R., & Tomescu, M. (2019). Renewable energy in Europe—2019. *European Environmental Agency-European Topic Centre on Climate Change Mitigation and Energy: Mol, Belgium.*

Cherubini, F., & Jungmeier, G. (2010). LCA of a biorefinery concept producing bioethanol, bioenergy, and chemicals from switchgrass. *The International Journal of Life Cycle Assessment, 15*, 53–66.

Chineke, T. C., & Ezike, F. M. (2010). Political will and collaboration for electric power reform through renewable energy in Africa. *Energy Policy, 38*(1), 678–684.

Colombo, E., Bologna, S., Masera, D., & others. (2013). *Renewable energy for unleashing sustainable development* (Vol. 390). Springer.

Cucchiella, F., & D'Adamo, I. (2013). Issue on supply chain of renewable energy. *Energy Conversion and Management, 76*, 774–780.

Dincer, I. (2000). Renewable energy and sustainable development: A crucial review. *Renewable and Sustainable Energy Reviews, 4*(2), 157–175.

Emenike, S. N., & Falcone, G. (2020). A review on energy supply chain resilience through optimization. *Renewable and Sustainable Energy Reviews, 134*, 110088.

Etcheverry, J. (2011). New climate protection, energy security, and employment creation strategies for Latin and North America based on renewable energy collaboration. *Latin American Policy, 2*(1), 43–57.

Fontes, C. H. O., Freires, F. G. M., & Saavedra, M. R. (2018). Sustainable and renewable energy supply chain: A system dynamics overview. *Renewable and Sustainable Energy Reviews, 82*, 247–259.

Francis, K., Dongying, S., Dennis, A., Edmund, N. N. K., & Kumah, N. Y. G. (2022). Network governance and renewable energy transition in sub-Saharan Africa: Contextual evidence from Ghana. *Energy for Sustainable Development, 69*, 202–210.

Gebara, C. H., & Laurent, A. (2023). National SDG-7 performance assessment to support achieving sustainable energy for all within planetary limits. *Renewable and Sustainable Energy Reviews, 173*, 112934.

Godil, D. I., Yu, Z., Sharif, A., Usman, R., & Khan, S. A. R. (2021). Investigate the role of technology innovation and renewable energy in reducing transport sector CO_2 emission in China: A path toward sustainable development. *Sustainable Development, 29*(4), 694–707.

Gold, S., & Seuring, S. (2011). Supply chain and logistics issues of bio-energy production. *Journal of Cleaner Production, 19*(1), 32–42.

González, M. O. A., Galvão, M. S., de Falani, S. Y. A., dos Santos Gonçalves, J., & da Silva, L. T. S. (2012). Open innovation practices in the development of wind energy supply chain: an exploratory analysis of the literature. *Product: Management and Development, 10*(2), 104–111.

Hassan, Q., Viktor, P., Al-Musawi, T. J., Ali, B. M., Algburi, S., Alzoubi, H. M., Al-Jiboory, A. K., Sameen, A. Z., Salman, H. M., & Jaszczur, M. (2024). The renewable energy role in the global energy Transformations. *Renewable Energy Focus, 48*, 100545.

Hiloidhari, M., Sharno, M. A., Baruah, D. C., & Bezbaruah, A. N. (2023). Green and sustainable biomass supply chain for environmental, social and economic benefits. *Biomass and Bioenergy, 175*, 106893.

Hirsh, H. S., Li, Y., Tan, D. H. S., Zhang, M., Zhao, E., & Meng, Y. S. (2020). Sodium-ion batteries paving the way for grid energy storage. *Advanced Energy Materials, 10*(32), 2001274.

Hoang, A. T., Nižetić, S., Olcer, A. I., Ong, H. C., Chen, W.-H., Chong, C. T., Thomas, S., Bandh, S. A., & Nguyen, X. P. (2021). Impacts of COVID-19 pandemic on the global energy system and the shift progress to renewable energy: Opportunities, challenges, and policy implications. *Energy Policy, 154*, 112322.

Holloway, R., Ho, D., Delotavo, C., Xie, W. Y., Rahimi, I., Nikoo, M. R., & Gandomi, A. H. (2023). Optimal location selection for a distributed hybrid renewable energy system in rural Western Australia: A data mining approach. *Energy Strategy Reviews, 50*, 101205.

Hossain, M. S., Madlool, N. A., Rahim, N. A., Selvaraj, J., Pandey, A. K., & Khan, A. F. (2016). Role of smart grid in renewable energy: An overview. *Renewable and Sustainable Energy Reviews, 60*, 1168–1184.

Iqbal, M., Azam, M., Naeem, M., Khwaja, A. S., & Anpalagan, A. (2014). Optimization classification, algorithms and tools for renewable energy: A review. *Renewable and Sustainable Energy Reviews, 39*, 640–654.

Irani, Z., Kamal, M. M., Sharif, A., & Love, P. E. D. (2017). Enabling sustainable energy futures: Factors influencing green supply chain collaboration. *Production Planning & Control, 28*(6–8), 684–705.

Kabel, T. S., & Bassim, M. (2020). Reasons for shifting and barriers to renewable energy: A literature review. *International Journal of Energy Economics and Policy, 10*(2), 89–94.

Khan, S. A. R., Qianli, D., SongBo, W., Zaman, K., & Zhang, Y. (2017). Environmental logistics performance indicators affecting per capita income and sectoral growth: Evidence from a panel of selected global ranked logistics countries. *Environmental Science and Pollution Research, 24*, 1518–1531.

Khan, S. A. R., Sharif, A., Golpîra, H., & Kumar, A. (2019). A green ideology in Asian emerging economies: From environmental policy and sustainable development. *Sustainable Development, 27*(6), 1063–1075.

Khan, S. A. R., Zhang, Y., Kumar, A., Zavadskas, E., & Streimikiene, D. (2020). Measuring the impact of renewable energy, public health expenditure, logistics, and environmental performance on sustainable economic growth. *Sustainable Development, 28*(4), 833–843.

Kim, K., Jeong, B., & Jung, H. (2014). Supply chain surplus: Comparing conventional and sustainable supply chains. *Flexible Services and Manufacturing Journal, 26*, 5–23.

Kushch, S., & Castrillo, F. P. (2017). A review of the applications of the block-chain technology in smart devices and distributed renewable energy grids. *ADCAIJ: Advances in Distributed Computing and Artificial Intelligence Journal, 6*(3), 75–84.

Li, Q., Loy-Benitez, J., Nam, K., Hwangbo, S., Rashidi, J., & Yoo, C. (2019). Sustainable and reliable design of reverse osmosis desalination with hybrid renewable energy systems through supply chain forecasting using recurrent neural networks. *Energy, 178*, 277–292.

Liu, Y., & Du, J. (2020). A multi criteria decision support framework for renewable energy storage technology selection. *Journal of Cleaner Production, 277*, 122183.

Marco-Lajara, B., Martínez-Falcó, J., Sánchez-García, E., & Millan-Tudela, L. A. (2023). Analyzing the role of renewable energy in meeting the sustainable development goals: A bibliometric analysis. *Energies, 16*(7), 3137.

McCollum, D. L., Echeverri, L. G., Busch, S., Pachauri, S., Parkinson, S., Rogelj, J., Krey, V., Minx, J. C., Nilsson, M., Stevance, A.-S., et al. (2018). Connecting the sustainable development goals by their energy inter-linkages. *Environmental Research Letters, 13*(3), 33006.

Mertens, J., Dewulf, J., Breyer, C., Belmans, R., Gendron, C., Geoffron, P., Goossens, L., Fischer, C., Du Fornel, E., Hayhoe, K., Hirose, K., Cadre-Loret, E. L., Lester, R., Maigné, F., Maitournam, H., de Miranda, P. E. V., Verwee, P., Sala, O., Webber, M., & Debackere, K. (2024). From emissions to resources: mitigating the critical raw material supply chain vulnerability of renewable energy technologies. *Mineral Economics*, 1–8.

Mirza, U. K., Ahmad, N., Harijan, K., & Majeed, T. (2009). Identifying and addressing barriers to renewable energy development in Pakistan. *Renewable and Sustainable Energy Reviews, 13*(4), 927–931.

Nematian, J., & Rahimi, I. (2022). Feasibility study of using renewable energies in Iranian Seas: A comparative study. *Renewable Energy, 189*, 383–391.

Nia, A. R., Awasthi, A., & Bhuiyan, N. (2021). Industry 4.0 and demand forecasting of the energy supply chain: A literature review. *Computers & Industrial Engineering, 154*, 107128.

Pacesila, M., Burcea, S. G., & Colesca, S. E. (2016). Analysis of renewable energies in European Union. *Renewable and Sustainable Energy Reviews, 56*, 156–170.

Potrč, S., Čuček, L., Martin, M., & Kravanja, Z. (2021). Sustainable renewable energy supply networks optimization—The gradual transition to a renewable energy system within the European Union by 2050. *Renewable and Sustainable Energy Reviews, 146*, 111186.

Rahimi, I., Nikoo, M. R., & Gandomi, A. H. (2023). Techno-economic analysis for using hybrid wind and solar energies in Australia. *Energy Strategy Reviews, 47*, 101092.

Raman, R., Sreenivasan, A., Ma, S., Patwardhan, A., & Nedungadi, P. (2023). Green Supply chain management research trends and linkages to UN sustainable development goals. *Sustainability, 15*(22), 15848.

Seuring, S., & Müller, M. (2008). From a literature review to a conceptual framework for sustainable supply chain management. *Journal of Cleaner Production, 16*(15), 1699–1710.
Shrestha, A., Mustafa, A. A., Htike, M. M., You, V., & Kakinaka, M. (2022). Evolution of energy mix in emerging countries: Modern renewable energy, traditional renewable energy, and non-renewable energy. *Renewable Energy, 199*, 419–432.
Wang, L., Yang, Y., Yang, D., & Zhou, Y. (2024). Role of supply chain disruptions and digitalization on renewable energy innovation: Evidence from G7 nations. *Energy Economics*, 108016.
Wu, X., Hu, X., Moura, S., Yin, X., & Pickert, V. (2016). Stochastic control of smart home energy management with plug-in electric vehicle battery energy storage and photovoltaic array. *Journal of Power Sources, 333*, 203–212.
Zhang, P., Sun, M., Zhang, X., & Gao, C. (2017). Who are leading the change? The impact of China's leading PV enterprises: A complex network analysis. *Applied Energy, 207*, 477–493.

Chapter 2
Energy Management in Renewable-Based Supply Chain with Considering Demand-Side Flexibility

Homa Rashidizadeh-Kermani, Miadreza Shafie-khah, and Pierluigi Siano

Abstract Despite the benefits of renewable resources in the network, uncertainties in renewable generation can impose multiple challenges in efficient operation of electricity supply chain in both supply and demand side. So, integration of distributed renewable generations in electricity markets through adaptive solutions to address the uncertainties should be investigated. In this regard, in this chapter, energy management of a renewable-based supply chain is investigated in which, in one layer, a wind power producer (WPP) offers strategically in electricity market while in the other layer, the DSO minimizes its operating costs. Also, to smooth the variations of the WPP, employing of demand response programs and transition of consumers into prosumers are also considered as efficient options. So, through the energy transactions in local electricity market, the WPP, DSO and loads are considered in a three-tier supply chain which is modeled as a bi-level problem in which the WPP maximizes its expected profit while the DSO minimizes its operating costs. By using Karush–Kuhn–Tucker optimality conditions the achieved problem is recast to a single-level one.

Keywords Distribution systems · Prosumer · Responsive load · Three-tier supply chain · Wind power producer

---

H. Rashidizadeh-Kermani (✉)
Department of Electrical and Computer Engineering, University of Birjand, Birjand, Iran
e-mail: rashidi_homa@birjand.ac.ir; rashidi.h@nowocert.com

H. Rashidizadeh-Kermani · M. Shafie-khah
Research and Innovation Division, Nowocert, Dublin, Ireland

P. Siano
Department of Management and Innovation Systems, University of Salerno, Salerno, Italy

I. Rahimi et al. (eds.), *Sustainable Supply Chain of Renewable Energy Networks*, Energy Systems in Electrical Engineering, https://doi.org/10.1007/978-981-96-4940-2_2

2.1 Introduction

The transformation of raw energy into usable energy which consists of significant sets of management stages from attainment of energy sources up to the consumption of that consumable energy is defined as renewable-based supply chain (Jelti et al., 2021). There exist significant parts for that kind of supply chain covering the input process which starts from raw materials to the output process as final product. The whole parts include procurement, generation, transmission, distribution and demand that can be totally divided into three main process including the upstream, production and downstream (Tsao & Thuy-Linh, 2019). With renewable-based supply chain, the use of renewable energy will be encouraged and the related technologies would be promoted. Among renewable resources, wind energy is a high potential resource that is currently available all around the world. On the other hand, electric grid faces a transition through integration of multiple renewable resources. This transition makes the imbalances between load and generation because of insufficient flexibility in the network. Therefore, in recent years, the demand side plays a critical role in supplying flexibility to the grid. So, demand response (DR) has the potential to reduce the peak load and so to mitigate the requirement for expanding the generation capacity (Rashidizadeh-Kermani et al., 2019). In (Guo et al., 2024), DR programs are implemented to reduce the total costs of the electric power supply chain through multiple pricing strategies. The global and local impact of uncertainties related to the distributed renewable generation units on the demand side in a supply chain through an active district is investigated in Yi et al. (2018). The accuracy for the matching supply and demand in a supply chain through digital economy is improved in Teng and Lin (2024). Existence of flexible and responsive resources can keep sustainability and resiliency in the energy supply chains that can avoid disruptions caused by pandemics such as COVID-19 (Kazancoglu & Ozbiltekin-Pala, 2022).

With increasing the proportion of renewable resources in trading electricity, the end-users behave as prosumers and have transformed their role from consumers to producers. In this context, some trading floors such as peer to peer (P2P) has been emerged enabling prosumers to send their excess electricity to their neighbors. Different opportunities for demand-side management are developed in Zheng et al. (2022) for commercial prosumers in Hong Kong to manage cooling and thermal loads. Generally, demand-side management plays a crucial role in power supply chain optimization in which coordination between the renewables and demand as well as managing the distribution system leads to reduce the costs and satisfying the customers preferences. In this regard, in Yue et al. (2023), an effective procedure for optimizing various types of energy infrastructure confronting with uncertain resources is proposed while guaranteeing the grid limitations. Energy storage system is used in Zhang et al. (2019) via a two-stage robust scheduling problem for keeping coordination between flexible loads and uncertainties of wind power generation to achieve efficient wind power consumption. These concerns have made an interest among the researches to focus on the activities in the field of integration of renewable resources with the electricity market. Multiple models have been proposed for modeling (Rashidizadeh-Kermani et al., 2022) and decarbonization of electricity markets (Yunfei et al., 2024). A hybrid supply chain is modeled in Ahmed et al. (2024) to combine a power plant and a grid-connected microgrid in which the storage systems absorbs the fluctuations in production capacity because of the abrupt weather changes natured from the renewable energy generation.

A wide range of models has investigated the interaction among renewable generation companies with the network with exercising the market power (Baringo & Concjo, 2013). A three-tier production supply chain network is implemented in Li et al. (2023) in which transactive energy plays a critical load in locating wind turbines. Extensive researches on optimizing renewable-based supply chain have been published in many publications. The manufacturing industries in Malaysia have developed the renewable energy resources in a supply chain to keep the trade-off between environmental problems and economic issues (Al-Madani et al., 2024). Optimal planning for renewable supply chains through optimization process is introduced in Al-Nory (2019) which supports electricity market decision-makers to determine optimal operating of a well-designed supply chain.

The intermittency of wind energy is reduced in conjunction of wind with pumped water (Yunfei et al., 2024), compressed air (Liu et al., 2024) and batteries in different stochastic optimization models. One of the most important challenges in the scheduling optimization problems is how to control the risks natured from the incidents which have harmful effect on utilization of supply chain. In different publications, various methods such as CVaR (Jelti et al., 2021), VaR (2024) and regret (Xu et al., 2017) have been combined to the problem to hedge against volatilities in scheduling problems. In (Rashidizadeh-Kermani et al., 2020), a risk averse approach is incorporated to decision-making problem to measure the worst-case scenarios. A two-stage stochastic programming model is presented in Bei et al. (2019) to achieve a recovery and strategic operational conditions in a supply chain. A stochastic model to control the risks associated with uncertainties of loads, wind power and grid accessibility is investigated in Zhu and Cao (2021) through a market-based framework.

In this chapter, the transactions among the decision-makers in a renewable-based power supply chain are investigated where it provides a framework to incorporate stochastic issues with risk control components. Within the proposed problem, the flexibility of responsive loads through implementing DR programs in supply chain process is also considered to supply the required demand in a local electricity trading floor. In addition, risk measurement tools are also introduced to control the variability seen in the problem.

This chapter is organized with the following sections:

In Sect. 2.2, a description to renewable-based supply chain is given. Strategic offering problem of WPP in renewable-based supply chain and the related optimization model are provided in Sects. 2.3 and 2.4, respectively. Risk management of the mentioned problem is also explained in Sect. 2.5. The barriers related to the development of renewable-based supply chain are discussed in Sect. 2.6. Lastly, the conclusion part is provided in Sect. 2.7.

2.2 Description of Renewable-Based Power Supply Chain

With growing the population, the global required energy demand increases. In this regard, insufficient supply from generation resources to meet the growing demand causes energy shortage. Fossil fuels that dominate the global electricity supply contribute to climate change, global warming and greenhouse gas emissions. To address this issue, renewable energy resources such as wind power have been developed. But, due to the intermittency of these resources, they cannot generate a continuous power supply for the loads. So, various options such as conventional units and demand-side management play an important role to keep balance between demand and supply side. In this regard, the supply chain problem explained here comprises multiple parts: First the WPP and other conventional units sell their generated power to the distribution system in day-ahead electricity market. The distribution system operator (DSO) purchases the electricity from the generation units in order to sell it to the loads. In real time, before the power delivery, the generation units should enter balancing market to compensate their power deviations. This problem is solved from the overview of the WPP with considering the goal of DSO and surely with implementing technical limitations. In addition, responsive loads would provide more flexible options for the DSO for scheduling the demand and generation resources to keep them in local electricity markets. In supply side, there exist government subsidies that help the renewable generations (here the WPP) to stay as a member of supply chain. The incentives and subsidies in demand side also help loads to participate in DR programs and so to remain in the supply chain.

Here, we introduce a three-tier renewable-based power supply chain as shown in Fig. 2.1 which consists of the following elements:

First Tier: generation units including the WPP and other DG units including conventional and other renewable generations.

Second Tier: the DSO who is responsible of distributing energy from generation resources to the final customers.

Third Tier: loads as responsive and non-responsive loads.

In the strategic model proposed here, all supply chain elements arrive to cooperate with each other through a competition to supply the total load. In this supply chain, various types of resources such as renewable generation, conventional unit and responsive and fix loads exist that are connected with each other through distribution network. Due to the existence of multiple uncertainties, coordination among the flexible elements should be implemented between the demand zone and the supply zone. In this supply chain model, the demand and supply zone are connected with each other through the distribution network and the generation units compete against each other to sell their electricity to the network through strategic policies. In addition, to hedge against the volatilities related to the supply chain logistics, some techniques are also implemented to reduce the system variability.

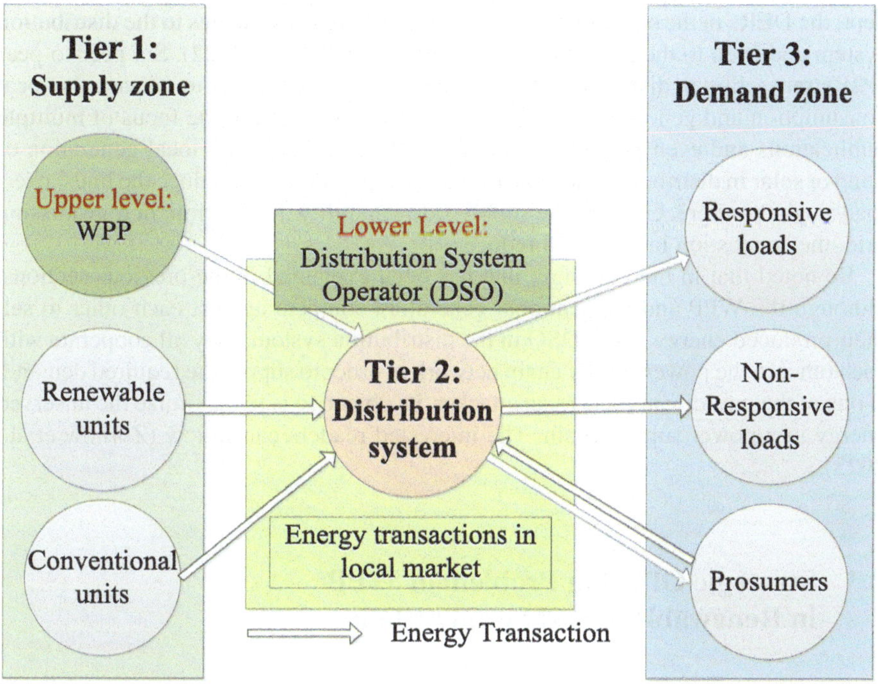

Fig. 2.1 Strategic offering problem of WPP in three-tier renewable-based power supply chain

2.2.1 Transactions in Renewable-Based Power Supply Chain Through a Local Electricity Market

In recent years, with communicational infrastructure, the electrical distribution network has changed from a passive grid to a responsive network in which the distributed generation units and also flexible responsive loads are able to adjust their production or consumption. In order to operate the smart grid efficiently, the coordination among the production and consumption systems to keep the balance between generation and demand looks necessary. In this regard, the transactive energy trading floor provides a local energy market environment for energy transactions in smart grids. So, with using market mechanisms the suppliers and consumers suggest their level of flexibility at a certain time and price to DSO.

In distribution side, an interconnection among different participants such as distributed energy resources (DERs), solar and wind power generations, electric vehicles and flexible responsive and non-responsive loads has been seen. Be noted that in demand zone, multiple residential, industrial and commercial customers may be equipped with local DERs and take a role of prosumer instead of just being a consumer. So, as seen in Fig. 2.1, the double side transactions will occur between the prosumers and other participants through the distribution system. If appropriately be kept, the DERs in the side of prosumers can make high advantages to the distribution system and even to the customers (MansourLakouraj et al., 2022). So, peer-to-peer (P2P) transaction in distribution side has been explored in order to manage the local consumption and generation. P2P electricity trading has been the focus of multiple publications and even projects (Liu et al., 2019). For example, local generation of wind or solar in distribution side can mitigate the power losses, reduce the lines' overloads and. therefore, enhance the supply side reliability. In addition, in transmission grid, the congestion in lines will reduce.

Be noted that in the paradigm that has been explained in the previous sections, although the WPP and conventional generators compete against each other to sell their produced energy to the DSO in the distribution system, they all cooperate with each other in the power supply chain network in order to supply the required demand. If one of the participants acts as an attacker, its objective is to maximize the unserved energy in a power supply chain. The interested readers can follow (Zornow et al., 2023).

2.3 Strategic Offering Problem of WPP in Renewable-Based Supply Chain

Here, we keep an electric power supply chain problem in which there exist two decision-makers including a renewable company (i.e., WPP) and the DSO who is responsible for the distribution network.

There are multiple generation units including conventional and renewable generation units who compete against each other to sell their generated energy to the DSO. In this model, unlike the conventional generation units, the WPP is considered with market power that tends to maximize its expected profit through its offering strategy. With this model, the WPP participates in day-ahead market and determines its volume of wind power and price that should be offered to that market. The WPP has an uncertain production and it has no control on it. So, it is obvious that the power delivery of the WPP will deviate from its scheduled production. To compensate this deviation, the WPP should take part in the balancing market to cover its surplus or deficit generation. In fact, through the balancing market, the WPP can handle the imbalances between the scheduling points and the actual power delivery. Be noted that the balancing market prices are unknown to the WPP and are subject to uncertainty.

From Fig. 2.1, it can be seen that the DSO purchases energy from all generation units and prosumers and sells it to the customers at distribution level through a local market. The customers include both fixed and responsive loads. The flexible loads who are elastic can be considered as aggregated residential, commercial or industrial loads, the aggregation of electric vehicles or even some prosumers who may have local generations. The aggregator on behalf of those loads can participate in the market to purchase their required demand from the upstream. Here, the demand of loads is also an uncertain parameter. In Fig. 2.1, there are some nodes in the supply chain that the connectivity among them is illustrated through a decision-making problem. Here, it is supposed that the WPP is the main decision-maker in the supply chain network where the decisions made by the DSO surely affect the WPP's intentions. So, the goal of the DSO should also be seen by the WPP. In this supply chain problem, the WPP is the main decision-maker who wants to maximize its profit, while the decisions made by the DSO also affects the decisions of the WPP. The DSO tends to minimize its operating costs and other supply chain members who want to sell their energy with high prices and purchase it with lower costs, respectively. This will be discussed in the next sections.

For modeling of this problem, there exist some assumptions including:

- All the producers including the conventional generators and the WPP can offer their generation to the DSO and not directly to the final customers.
- All the generation units are paid in day-ahead market with the price at each node.
- The WPP as a decision-maker acts strategically with its market power only in day-ahead market, although this strategic behavior can be expanded to the balancing market also.
- Only the WPP has the market power while the other market participants compete in the market ideally.

Here, there exists a strategic WPP who wants to offer in day-ahead market and seeks to maximize its expected profit, while it should compensate its power deviations through energy transactions in the balancing market. This problem is constrained with a market clearing problem that is provided as an optimization problem in which an agent-based decision-making problem (for the WPP) is considered in the upper level which is subject to the other decision-making problem (for the DSO) in the lower level. It should be mentioned that the market clearing problem in the lower level is surely affected by the actions of the strategic WPP in the upper level.

From the above explanations, it should be also noted that this problem consists of two stages, where in the first stage the WPP determines its volume of offering generation and the related prices in the day-ahead market. In the next stage, the WPP compensates its excessive/deficit of production in the balancing market.

To make the problem solvable, the lower level problem is substituted with its Karush–Kuhn–Tucker (KKT) optimality conditions. So, the model will be recast to a single-level problem that can be solved using commercial solvers. Figure 2.2 depicts the process that is applied to the problem to solve it.

2.3.1 Problem Uncertainties

There are multiple types of uncertainties in this problem including the balancing market prices, wind power generation, demand of loads and bidding prices of loads. Other sources of uncertainty such as the photovoltaic generation of prosumers can also be added to the set of the uncertainties.

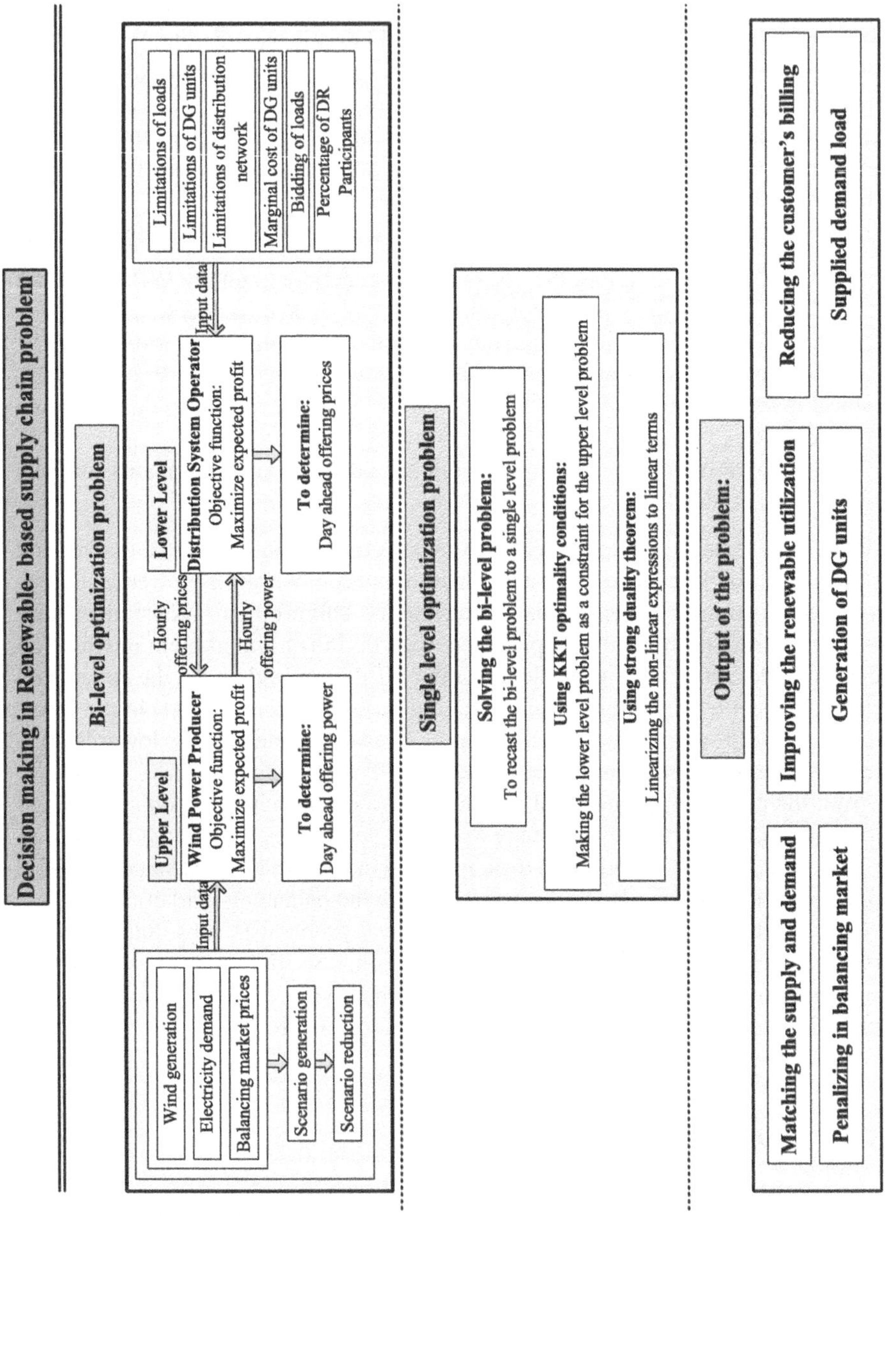

Fig. 2.2 Solution methodology of decision-making in renewable-based supply chain problem

To model the problem, these uncertainties are provided through different scenarios as input to the problem. The related scenarios can be generated using time series (Gabriel et al., 2012) or with using their probability density function (PDF) (Conejo et al., 2010). The generated scenarios which are used to model the random variable are arranged via a scenario tree. A scenario tree has a set of arcs and nodes. Each node represents the points that the decisions are made. The root of this tree is the first node which stands for the decisions made in the commence of the first stage of the planning horizon. The second stage decisions are made in the next nodes which are connected to the root ones. The last nodes represent the final stage of the planning horizon called as leaves of the tree. In such a scenario tree, each single trajectory between each root node and each leave is called scenario. In addition, the arcs demonstrate the realization of random variables with the related occurrence probability (Gabriel et al., 2012). Be noted that some of the scenarios may be related to each other. For example, the scenarios of wind power generation can be related to the photovoltaic power generation. So, to reduce the number of scenarios, some of the uncertainty resources are considered connected to each other. However, with increasing the number of scenarios, k-means is a popular clustering algorithm that is applied which assigns data points to the centroids (Nguyen et al., 2018). There are two main tasks in k-means algorithm including: determining K centers for centroids that is often chosen with iterative processes and assigning the data points to the selected K centroids and creating a cluster of data. Be noted that each cluster of data points has common characteristics that is away from other clusters. Euclidean distance is a usual metric that is generally used to assess the distance between each point and the centroid of the clusters. Each data point is associated to the closest centroid which has the smallest distance with it. The Euclidean distance corresponds to the shortest geometric distance between two pints of x_i and y_i as $d = \sqrt{\sum_{i=1}^{N} (x_i - y_i)^2}$. Finally, the reduced number of scenarios from k-means algorithm is given as input to the program. A priori for the number of scenarios is not known generally; however, few number of scenarios is desirable because each scenario will be added to the stochastic program which can significantly affect the tractability of solving the program in a reasonable time.

2.3.2 *Modeling Responsive Loads in Renewable-Based Supply Chain*

Responsiveness as a critical link between suppliers and customers has been an interesting topic for the managers in renewable-based supply chains that brings market sharing and financial performance. With applying DR programs, the customers adjust their energy consumption based on the electricity price signals. So, the pattern of their consumption will change from an initial value of $P_{t,\omega}^{D,\text{int}}$ to $P_{t,\omega}^{D}$. The index of t, ω denotes at each time t and scenario ω.

$$P_{t,\omega}^{T_D} = P_t^{D,\text{int}_{t,\omega}^{D}}. \tag{2.1}$$

With tuning the consumption behavior of electricity energy of end-users, the benefit of customers is obtained as:

$$B\left(P_{t,\omega}^{T_D}\right) = S\left(P_{t,\omega}^{T_D}\right) - P_{t,\omega}^{T_D}\rho_{t,\omega}^{DA}, \tag{2.2}$$

where $B\left(P_{t,\omega}^{T_D}\right)$ and $S\left(P_{t,\omega}^{T_D}\right)$ denote the benefit and income of customers at period t after implementing DR programs, respectively. In order to maximize the benefit of end-users, the expression in (2.3) should be satisfied.

$$\frac{\partial B\left(P_{t,\omega}^{T_D}\right)}{\partial P_{t,\omega}^{T_D}} = 0 \Rightarrow \frac{\partial S\left(P_{t,\omega}^{T_D}\right)}{\partial P_{t,\omega}^{T_D}} - \rho_{t,\omega}^{DA} = 0. \tag{2.3}$$

In order to model the participation of responsive loads in DR programs, a quadratic function is taken as:

$$B\left(P_{t,\omega}^{T_D}\right) = B_t^{D,\text{int}} + \frac{\rho_h^{DA,\text{int}} P_{t,\omega}^{T_D}}{1+\text{Elas}_{t,h}^{-1}} \times \left[\left(\frac{P_{t,\omega}^{T_D}}{P_t^{D,\text{int}}}^{\text{Elas}_{t,h}^{-1}}\right) - 1\right]. \tag{2.4}$$

With taking differentiate from (2.4) with respect to:

$$\frac{\partial B\left(P_{t,\omega}^{T_D}\right)}{\partial P_{t,\omega}^{T_D}} = $$

$$\frac{\rho_t^{DA,\text{int}}}{1+\text{Elas}_{t,h}^{-1}} \times \left[\left(\frac{P_{t,\omega}^{T_D}}{P_t^{D,\text{int}}}^{\text{Elas}_{t,h}^{-1}} - 1\right)\right] + \frac{\rho_h^{DA,\text{int}_{t,\omega}^{T_D}}}{1+\text{Elas}_{t,h}^{-1}}$$

$$\times \left[\text{Elas}_{t,h}^{-1} \times \frac{1}{P_t^{D,\text{int}}}\left(\frac{P_{t,\omega}^{T_D}}{P_h^{D,\text{int}}}\right)^{\text{Elas}_{t,h}^{-1}}\right] \tag{2.5}$$

Let's substitute (2.5) into (2.3):

$$\left(1+\text{Elas}_{t,h}^{-1}\right) \times \frac{\rho_h^{DA,int}}{\rho_{t,\omega}^{DA}} = \left(\frac{P_{t,\omega}^{T_D}}{P_t^{D,int}}\right)^{\text{Elas}_{t,h}^{-1}} - 1 + \text{Elas}_{t,h}^{-1}\left(\frac{P_{t,\omega}^{T_D}}{P_h^{D,int}}\right)^{Elas_{t,h}^{-1}} \tag{2.6}$$

$$\frac{\rho_{h,\omega}^{DA}}{\rho_h^{int}} = \left(\frac{P_{t,\omega}^{T_D}}{P_t^{D,int}}\right)^{\text{Elas}_{t,h}^{-1}} - \frac{1}{1+\text{Elas}_{t,h}^{-1}}. \tag{2.7}$$

That we can have:

$$P_{t,\omega}^{T_D} = P_t^{D,init}\left(\frac{\rho_{t,\omega}^{DA}}{\rho_t^{DA,int}} + \frac{1}{1+\text{Elas}_{t,h}^{-1}}\right). \tag{2.8}$$

The price elasticity can be given in multiple references such as (Vahedipour-Dahraie et al., 2017) which indicates the behavior of customers to the variation of electricity prices. Here, the mathematical model of flexible loads is given as:

$$P_{t,\omega}^{T_D} = P_{t,\omega}^{D,\text{int}} \exp\sum_{h=1}^{T} \text{Elas}_{t,h}\text{Ln}\left[\frac{\rho_{t,\omega}^{DA}}{\rho_{t,\omega}^{DA,\text{int}}} + \frac{1}{1+\text{Elas}_{t,h}^{-1}}\right]. \tag{2.9}$$

In the above expressions, $P_{t,\omega}^{D,\text{int}}$ shows the initial demand before implementing DR programs while $P_{t,\omega}^{T_D}$ denotes the requested demand after DR programs.

2.4 Optimization Problem of Strategic Offering of WPP in Renewable-Based Supply Chain

As explained before, the offering problem of a strategic WPP is formulated as a bi-level model in which the WPP in the upper level maximizes its expected profit, while the DSO in the lower level minimizes its expected costs. Here, let $P_{t,\omega}^{cl}$ and $\rho_{n,t,\omega}^{DA}$ denote the cleared wind power and the locational marginal pricing (LMP) at bus n in day-ahead market, $P_{t,\omega}^{P}$ and ρ_t^P show the forecasted power and the marginal cost of WPP, and $P_{t,\omega}^{\text{pos/neg}}$ and $\rho_{t,\omega}^{\text{pos/neg}}$ represent the transacted power and the related prices in balancing market. π_ω is the probability at scenario ω.

$$\text{Max}\sum_{\omega\in\Omega}\pi_\omega\sum_{t\in T}\left[P_{t,\omega}^{cl}\rho_{t,\omega}^{DA} - P_{t,\omega}^{P}\rho_t^{P} + P_{t,\omega}^{pos}\rho_{t,\omega}^{pos} - P_{t,\omega}^{neg}\rho_{t,\omega}^{neg}\right] \tag{2.10}$$

The expression in (Eq. 2.10) defines the expected profit of the WPP which include the cleared energy in day-ahead market, the forecasted wind power and the last two terms illustrate the balancing power sold/ purchased due to the excessive/deficit of energy in the balancing market. For the above formulation, there exists a balancing constraint as below:

$$P_{t,\omega}^{P} - P_{t,\omega}^{pos} + P_{t,\omega}^{neg} = P_{t,\omega}^{cl}. \tag{2.11}$$

Equation (2.11) claims that the deviations of the forecasted and the cleared wind power should be covered in the balancing market. The result of that is equal with the wind power cleared in the day-ahead market. It should be mentioned that the forecasted power of the WPP is restricted within its capacity and also the cleared power of the WPP is less than its forecasted power that are shown in the following constraints:

$$P^{P}_{t,\omega} \le \overline{P}^{WPP}, \tag{2.12}$$

$$P^{cl}_{t,\omega} \le P^{P}_{t,\omega}. \tag{2.13}$$

The upper-level problem is subject to the lower-level problem which is about minimizing the operating costs of the DSO. The lower-level problem is given as follows. Let $P^{G}_{t,\omega}$ and ρ^{G} denote the scheduled power and marginal costs related to the conventional generators. $\alpha_{W,t}$ is the offering price of the WPP. The generated power from the demand side as $P^{P_D}_{t,\omega}$ is also purchased from the prosumers. $P^{D}_{t,\omega}$ and $P^{T_D}_{t,\omega}$ show the non-responsive and elastic loads with the price signal of $\rho^{D}_{t,\omega}$.

$$\text{Min} \sum_{\omega \in \Omega} \pi_\omega \sum_{t \in T} \left[P^{G}_{t,\omega}\rho^{G} + \alpha_{W,t}P^{cl}_{t,\omega} - \left(P^{D}_{t,\omega} + P^{T_D}_{t,\omega}\right)\rho^{D}_{t,\omega} + P^{P_D}_{t,\omega}\rho^{D}_{t,\omega} \right]. \tag{2.14}$$

In this problem, the DSO minimizes its operating costs to determine the LMPS. The LMPs can be achieved by solving either the ac optimal power flow (ACOPF) or dc optimal power flow (DCOPF) (Rashidizadeh-Kermani et al., 2023). Here, we use DCOPF due to its sufficient mathematical calculation and approximation accuracy (Li et al., 2014). The objective of the DSO is subject to different constraints that are provided as follow:

The power balance at each node is described with Eq. (2.15) that explains the power transaction at each node equals with the power generation of both DG units and WPP and the produced power by the prosumers minus the total responsive and non-responsive loads at that node.

$$\sum_{k|s(k)=n} f_{k,t,\omega} + \sum_{k|r(k)=n} f_{k,t,\omega} = \sum_{G \in NG} P^{G}_{t,\omega} + P^{cl}_{t,\omega} + P^{P_D}_{t,\omega} - P^{D}_{t,\omega} - P^{T_D}_{t,\omega} : \rho^{DA}_{n,t,\omega}. \tag{2.15}$$

The power flow at each branch is given by Eq. (2.16) which is limited within is limitation as in (2.17).

$$f_{k,t,\omega} = B_k\left(\delta_{s(k,\omega)} - \delta_{r(k,\omega)}\right) : \beta_k \tag{2.16}$$

$$-f_k^{\max^{\max}_{k,t,\omega,k}} : \beta_k^{\min^{\max}_k}. \tag{2.17}$$

The restrictions related to the conventional generation units and also the limitations of the demand of loads are also given here, respectively:

$$0 \le P^{D}_{t,\omega} \le \overline{P}^{\max}_{D} : \varepsilon_D^{\min^{\max}_D}, \tag{2.18}$$

$$0 \le P^{T_D}_{t,\omega} \le \overline{P}^{T_D,\max} : \varepsilon_{T_D}^{\min^{\max}_{T_D}}, \tag{2.19}$$

$$0 \leq P_{t,\omega}^{P_D} \leq \overline{P}^{P_D,\max} : \varepsilon_{P_D}^{\min_{P_D}^{\max}}, \tag{2.20}$$

$$0 \leq P_{t,\omega}^{G} \leq \overline{P}^{G} : \varepsilon_{G}^{\min_{G}^{\max}}, \tag{2.21}$$

$\overline{P}_D^{\max}, \overline{P}^{T_D,\max}$ are the maximum amount of non-responsive and responsive loads. $\overline{P}^{P_D,\max}$ and $\overline{P}_{t,\omega}^{G}$ are the maximum amount of power generated by the DG units and the prosumers.

At each node, the angles are limited within its bound as in (2.22) while the slack bus in (2.23) should not be forgotten!

$$-\pi \leq \delta_{n,t,\omega} \leq \pi \forall n \backslash n : \text{ref} : \mu_n^{\min_n^{\max}}, \tag{2.22}$$

$$\delta_{n,t,\omega} = 0\, n : \text{ref} : \gamma_n. \tag{2.23}$$

The variables that are separated with colon in front of each constraint represent the dual variables in the bi-level problem.

2.4.1 Optimization Solution for Strategic Offering Problem in Renewable-Based Supply Chain

Here, we have a bi-level optimization problem with a specific objective function in each level. To solve the problem, the bi-level problem should be converted to a single-level one. So, the lower level problem can be replaced with its Karush–Kuhn–Tucker optimality conditions. To obtain KKT conditions, the Lagrangian function of the lower level problem is obtained with respect to its objective functions, equalities and non-equality constraints.

To explain briefly, for an objective function such as:

$$\begin{aligned} &\text{Minimize}_x f(x); \\ &\text{subject to :} \\ &\{h(x) = 0 \text{ and } g(x) \leq 0\} \end{aligned} \tag{2.24}$$

The KKT optimality conditions should be satisfied. To obtain the KKT conditions, the Lagrangian function is defined as:

$$L = f(x) + \lambda_T h(x) + \mu_T g(x), \tag{2.25}$$

where $f(x)$, $h(x)$ and $g(x)$ are continuously differentiable in feasible regions as: $x \in \{x|h(x) = 0, g(x) \leq 0\}$. So, KKT conditions will be defined as:

$$\nabla_x f(x) + \lambda_T \nabla_x h(x) + \mu_T \nabla_x g(x) = 0, \tag{2.26}$$

where

$$h(x) = 0, g(x) \leq 0, \mu_T g(x) = 0, \mu \geq 0. \tag{2.27}$$

Above, λ and μ are the vectors of Lagrange multipliers related to equalities and inequalities. ∇_x refers to the gradient with respect to the variable x. The equation in (2.26) denotes the Lagrangian function equals with zero while the conditions in (2.27) should be satisfied. In this regard, the gradient of the Lagrangian function of our problem with respect to the optimization variables are given as below:

$$\rho^G - \rho_{n,t,\omega}^{DA} - \varepsilon_G^{\min_G^{\max}}, \tag{2.28}$$

$$\alpha_{W,t} - \rho_{n,t,\omega}^{DA} - \varepsilon_W^{\min_W^{\max}}, \tag{2.29}$$

$$\rho_{t,\omega}^D - \rho_{n,t,\omega}^{DA} - \varepsilon_D^{\min_D^{\max}}, \tag{2.30}$$

$$\rho_{t,\omega}^D - \rho_{n,t,\omega}^{DA} - \varepsilon_{T_D}^{\min_{T_D}^{\max}}, \tag{2.31}$$

$$\rho_{t,\omega}^D - \rho_{n,t,\omega}^{DA} - \varepsilon_{P_D}^{\min_{P_D}^{\max}}, \tag{2.32}$$

$$\rho_{n(s),t,\omega}^{DA} - \rho_{n(r),t,\omega}^{DA} - \beta_k - \beta_k^{\min_k^{\max}}, \tag{2.33}$$

$$\sum_{k|s(k)=n} B_k \beta_k - \sum_{k|r(k)=n} B_k \beta_k - \mu_n^{\min_n^{\max}}, \tag{2.34}$$

$$\sum_{k|s(k)=n} B_k \beta_k - \sum_{k|r(k)=n} B_k \beta_k - \gamma_n = 0,\ n\ :\text{ref}. \tag{2.35}$$

The obtained problem may include some nonlinearities which should be replaced with the linear expressions. Here, using strong duality theory and mathematical relaxations, the linear form of the single-level problem would be obtained that can be solved using commercial solvers. The constraints in (2.36–2.49) are the complementarity conditions that are given here:

$$0 \leq \beta_k^{\max_k \max_{k,t,\omega}}, \tag{2.36}$$

$$0 \leq \beta_k^{\min_{k,t,\omega k}^{\max}}, \tag{2.37}$$

$$0 \leq \varepsilon_D^{ma\overline{x}_D^{\max_{t,\omega}^D}}, \quad (2.38)$$

$$0 \leq \varepsilon_D^{\min_{t,\omega}^D}, \quad (2.39)$$

$$0 \leq \varepsilon_{T_D}^{ma\overline{x}^{T_D,\max_{t,\omega}^{T_D}}} \quad (2.40)$$

$$0 \leq \varepsilon_{T_D}^{\min_{t,\omega} T_D}, \quad (2.41)$$

$$0 \leq \varepsilon_{P_D}^{ma\overline{x}^{P_D,\max_{t,\omega}^{P_D}}}, \quad (2.42)$$

$$0 \leq \varepsilon_{P_D}^{\min_{t,\omega} P_D}, \quad (2.43)$$

$$0 \leq \varepsilon_G^{\min_{t,\omega}^G}, \quad (2.44)$$

$$0 \leq \varepsilon_G^{ma\overline{x}_G^{\max_{t,\omega}^G}}, \quad (2.45)$$

$$0 \leq \varepsilon_W^{\max_{t,\omega t,\omega}^{off\ cl}}, \quad (2.46)$$

$$0 \leq \varepsilon_W^{\min_{t,\omega}^{cl}}, \quad (2.47)$$

$$0 \leq \mu_{n,t,\omega}^{\max_{n,t,\omega}}, \quad (2.48)$$

$$0 \leq \mu_{n,t,\omega}^{\min_{n,t,\omega}}, \quad (2.49)$$

The sign $\perp$ indicates complementarity. For the above equations for the variables x, y, the expression $0 \leq x \perp y \geq 0$ means that $x.y = 0,\ x \geq 0\,,\ y \geq 0$. The above nonlinear terms can be replaced with their related expressions using Fortuny-Amat transformation in Vahedipour-Dahraie et al. (2018).

In the above expressions, the production of $P_{t,\omega}^{cl}\rho_{t,\omega}^{DA}$ is nonlinear. Based on the convexity of the problem, with using strong duality theory and some mathematical relations (Fortuny-Amat & McCarl, 1981), this bilinear expression can be converted to a linear form:

$$P_{t,\omega}^{cl}\rho_{t,\omega}^{DA} = \sum_{D \in N_D} (P_{t,\omega}^D + P_{t,\omega}^{T,D} - P_{t,\omega}^{P_D})\rho_{t,\omega}^D - \sum_{G \in N_G} P_{t,\omega}^G \rho_t^G - \sum_{t \int N_L} f_l^{\max}(\beta_l^{\min} + \beta_l^{\max})$$
$$+ \sum_{G \in N_G} \left(-\varepsilon_G^{\max}\overline{P}^G - \varepsilon_D^{\max}\overline{P}_D^{\max} - \varepsilon_D^{\max}\overline{P}^{T_D,\max} - \varepsilon_{P_D}^{\max}\overline{P}^{P_D,\max} - \varepsilon_W^{\max} P_{t,\omega}^{off} \right)$$

$$-\sum_{n} \pi \left(\mu_n^{\min} + \mu_n^{\max}\right) \tag{2.50}$$

In the above problem, real data can be imported to be analyzed; for example, the balancing market prices can be extracted from Nordpool. The load and the wind power can be forecasted with mathematical methods such as time series, probability density function and artificial intelligence methods. The run time of this program is less than fifteen minutes over the horizon of 24 h which shows that this program can be used for practical implications. The practical data can be given to the interested reader if requested.

2.5 RISK Management in Renewable-Based Supply Chain

To control the risk natured from uncertain resources of our problem such as loads, market prices and wind power generation, different methods can be applied to the problem. For example, in Xu et al. (2017), regret gives the losses in the profit of a WPP that is caused by imperfect information from the prediction of wind energy prices. The regret corresponding with each scenario can be determined with the difference between the value of the objective function that is obtained based on perfect information and the value of the objective function under imperfect conditions. For example, the difference between the expected profit of the WPP under perfect information of wind power and the decisions made by the WPP when realizing the scenarios of wind power in the conditions of incomplete information from the WPP can be defined as the regret. Another method to measure financial losses in a portfolio specifically during a time frame is value at risk (VaR); however, one important disadvantage of VaR is that it provides no information about the distribution of profit (the objective) beyond its value. So, the fat tail appearing in the profit distribution is not determined in VaR method.

Supply chains in the business environment often confront with risks due to the complexity and their nature behavior. In a supply chain, all the decision-makers should use some risk aversion tools to choose the suitable strategies to assess their supply chain network. The supply chain in this chapter contains a problem in which one main source of uncertainties is electricity market prices. As briefly discussed in Gabriel et al. (2012), CVaR is the most common measurement tool to control the risks revealed in electricity markets. In our bi-level problem, the decision-maker who is the WPP should identify the combination of risk mitigation with the problem and keep its expected profit acceptable. In this supply chain risk management problem, with using conditional value at risk (CVaR), the WPP can choose multiple tactics and policies to choose its risk profile. CVaR is mathematically added to the problem with the following formulation:

$$\text{CVaR} = \max_{\xi,\eta_\omega}\left(\xi - \frac{1}{1-\alpha}\sum_{\omega=1}^{\Omega}\pi_\omega \cdot \eta_\omega\right). \tag{2.51}$$

Subject to:

$$\eta_\omega + \text{profit}_\omega - \xi \geq 0;\ \eta_\omega \geq 0, \tag{2.52}$$

where profit_ω stands for the profit in scenario ω; here, profit_ω is the profit of the WPP. η_ω as an auxiliary non-negative variable states the difference between auxiliary variable ξ and the profit_ω when the profit_ω is smaller than ξ and π_ω is the probability of scenario ω. Through a sensitive analysis, strategic offering problem of the WPP will be solved with different risk preferences in a supply chain problem.

CVaR term should be added to the formulation in (2.10) through a coefficient as β. When the WPP is risk-neutral, it tends to maximize its expected profit; so β should be chosen equal with zero. But, when the WPP behaves as a risk averse decision-maker, it is required to incorporate the risk measurement tool of CVaR to the problem with high values of β.

2.6 Developments of Renewable Energy Supply Chain: Benefits and Barriers!

For promotion of renewable energy resources, there exist multiple problems that can be totally separated into four categories including political, economic, managerial and technical barriers.

The integration of renewable energy resources with the power system confronts with different problems in improving the policies and regulations. Transition from fossil fuels to renewable energy resources requires skillful expertise with practical experiences with hands-on work in the related fields.

The other problem in front of the development of renewables is the high investment costs compared with fossil fuels which causes an increment in the electricity costs (Li & Bo, 2007). Likewise, the operation, maintenance, storage systems and also other expenses such as transportation take high costs in deployment of renewable resources (Rashidizadeh-Kermani et al., 2022). To reduce these expenses, governmental subsidies and incentives are very impressive which are unfortunately difficult to obtain due to major obstacles and limitations (Thanki & Thakkar, 2018). In addition to these obstacles, the lack of communication among the operators in the power systems makes limitations for expansion of the supply chain of renewable resources. On the other hand, sophisticated infrastructure of power system sectors in the absence of connection among government entities with delays in implementation of renewable energy projects makes the promotion of renewable-based supply chain suffer greatly (Zhang & Guo, 2024). Multiple platforms such as Internet of Things (IOT) through automated equipment and intelligent network lead to increase the efficiency within the prediction analysis and big data technologies (Lehyani & Zouari, 2015). With IOT, various performances such as manufacturing, services and constructions would be reshaped specifically in industrial centers through monitoring (Sree Krishna Das, 2023). Based on IOT, digitalization of renewable supply chains has been investigated in Rejeb et al. (2022) to augment the sustainability of renewable resources. In (Masoomi, 2024), blockchain technology has been implemented through a supply chain to adopt the traceability and pricing decisions between buyers and sellers.

From technical overview, there are major drawbacks in variable input and output of renewable energy technologies. For example, for photovoltaic systems, several factors such as humid and cloudy weather affect the efficiency of solar panels because of the reduction in absorbing the sun's rays. Also, with increasing the temperature, the heat absorption reduces the efficiency of semiconductors (Awasthy et al., 2024). The other important technical drawbacks in implementing renewable projects relate to their intermittency that causes high costs to the power system for keeping power balance during power delivery in real time (Ahmad & Zabri, 2018). For example, during low production of renewables and high demand, the grid operator should commit the generators such that to adjust the output generation of renewable resources to compensate the fluctuations (Westphal, 2024).

2.7 Conclusion

In this chapter, an offering strategy for a WPP in an oligopoly competition based on three-tier supply chain problem is introduced. The investigated problem is modeled as a bi-level model in which the WPP acts as a main decision-maker in the first level that should consider the behavior of the DSO who is in the lower level. This problem is recast to a single-level problem using mathematical methods. In the proposed framework, DR programs are employed to smooth the power variations. So, by implementing DR programs, customers that are equipped with DERs and have the role of prosumers are motivated to act with flexibility and respond to electricity price signals specifically during supply shortages. Here, responsive loads are considered to have a joint cooperation with renewable generations; this causes to improve not only the economy of all participants but also to reduce the dependency on the volatile upward markets.

In the described renewable-based supply chain profile, there exist multiple uncertainty resources such as market prices, loads and wind generation. To control the risks associated with the tiers of the under-study supply chain, the problem is formulated as a stochastic programming model and CVaR is added to it. So, integration of supply chain problem with risk measurement tools under multiple risk aversion factors leads to profitable and reliable operation of the model.

References

Ahmad, K., & Zabri, S. M. (2018). The deployment of performance measurement system under the supply chain management environment: The case of Malaysian manufacturing companies. *Management Production Engineering Review, 9*, 3–12. (In English).

Ahmed, W., Salman, S., Imran, M., Agha, M. H., & Moazzam, M. (2024) An inventory-based integrated model of a hybrid electricity supply chain for sustainable electricity storage management. *Journal of Energy Storage, 83*, 110746.

Al-Madanı, M. H. M., Fernando, Y., Iranmanesh, M., Lim, M. K., & Tseng, M.-L. (2024). Renewable energy supply chain in Malaysia: Fostering energy management practices and ecological performance. *Renewable Energy, 226*, Article 120441.

Al-Nory, M. T. (2019). Optimal decision guidance for the electricity supply chain integration with renewable energy: Aligning smart cities research with sustainable development goals. *IEEE Access, 7*, 74996–75006.

Awasthy, P., Haldar, T., & Ghosh, D. (2024). Blockchain enabled traceability—An analysis of pricing and traceability effort decisions in supply chains. *European Journal of Operational Research.*

Baringo, L., & Conejo, A. J. (2013). Strategic offering for a wind power producer. *IEEE Transactions on Power Systems, 28*(4), 4645–4654.

Bei, X., Zhu, X., & Coit, D. W. (2019). A risk-averse stochastic program for integrated system design and preventive maintenance planning. *European Journal of Operational Research, 276*(2), 536–548.

Liu, C., Su, X., Yin, Z., Sheng, Y., Zhou, X., Xu, Y., Wang, X., & Chen, H. (2024). Experimental study on the feasibility of isobaric compressed air energy storage as wind power side energy storage. *Applied Energy, 364*, 123129.

Conejo, A. J., Carrion, M., & Morales, J. M. (2010). *Decision making under uncertainty in electricity markets*. Springer.

Fortuny-Amat, J., & McCarl, B. (1981). A representation and economic interpretation of a two-level programming problem. *The Journal of the Operational Research Society, 32*(9), 783–792.

Gabriel, S. A., Conejo, A. J., David Fuller, J., Hobbs, B. F., & Ruiz, C. (2012). *Complementarity modeling in energy markets*, Springer.

Guo, Y., Liu, F., Song, J.-S., & Wang, S. (2024). Supply chain resilience: A review from the inventory management perspective. *Fundamental Research.*

Jelti, F., Allouhi, A., Büker, M. S., Saadani, R., & Jamil, A. (2021). Renewable power generation: A supply chain perspective. *Sustainability, 13*(3), 1271.

Kazancoglu, I., Ozbiltekin-Pala, M., Mangla, S. K., Kazancoglu, Y., & Jabeen, F. (2022). Role of flexibility, agility and responsiveness for sustainable supply chain resilience during COVID-19. *Journal of Cleaner Production, 362*, 132431.

Lehyani, F., & Zouari, A. (2015). Towards modeling the supply chain's performance evaluation criteria. Presented at the First International Conference on Transportation and Logistics (ICTL 2015). Sousse, Tunisia, 13 May 2015; Available online: https://hal.archivesouvertes.fr/hal-01893165 (Accessed on 21 April 2020).

Li, F., & Bo, R. (2007). DCOPF-based LMP simulation: Algorithm, comparison with ACOPF, and sensitivity. *IEEE Transactions on Power Systems, 22*(4), 1475–1485.

Li, R., Wu, Q., & Oren, S. S. (2014). Distribution locational marginal pricing for optimal electric vehicle charging management. *IEEE Transactions on Power Systems, 29*(1), 203–211.

Li, H., Jin, T., & Novoa, C. (2023). Facility and microgrid location-allocation for integrated supply chain and transactive energy operations. *Applied Mathematical Modelling, 119*, 119–136.

Liu, Y., Lei, W., & Li, J. (2019). Peer-to-peer (P2P) electricity trading in distribution systems of the future. *The Electricity Journal, 32*(4), 2–6.

MansourLakouraj, M., Shahabi, M., Shafie-khah, M., & Catalão, J. P.S. (2022). Optimal market-based operation of microgrid with the integration of wind turbines, energy storage system and demand response resources. *Energy, 239*(Part B), 122156.

Masoomi, B., Sahebi, I. G., Gholian-Jouybari, F., Mejia-Argueta, C., & Hajiaghaei-Keshteli, M. (2024). The role of internet of things adoption on the sustainability performance of the renewable energy supply chain: A conceptual framework. *Renewable and Sustainable Energy Reviews, 202*, 114610.

Nguyen, H. T., Le, L. B., & Wang, Z. (2018). A bidding strategy for virtual power plants with the intraday demand response exchange market using the stochastic programming. *IEEE Transactions on Industry Applications, 54*(4), 3044–3055.

Rashidizadeh-Kermani, H., Vahedipour-Dahraie, M., Shafie-Khah, M., & Siano, P. (2020). A regret-based stochastic bi-level framework for scheduling of DR aggregator under uncertainties. *IEEE Transactions on Smart Grid, 11*(4), 3171–3184.

Rashidizadeh-Kermani, H., Vahedipour-Dahraie, M., Parente, M., Shafie-Khah, M., & Siano, P. (2022). Strategic offering of a price maker wind power producer in distribution-level energy markets in presence of flexible prosumers. *IEEE Access, 10*, 21475–21485.

Rashidizadeh-Kermani, H., Vahedipour-Dahraie, M., Shafie-khah, M., & Siano, P. (2022). Evaluating the impact of bilateral contracts on the offering strategy of a price maker wind power producer. *IEEE Transactions on Industrial Informatics, 18*(7), 4331–4341.

Rashidizadeh-Kermani, H., Vahedipour-Dahraie, M., Shafie-khah, M., & Siano, P. (2022). A peer-to-peer energy trading framework for wind power producers with load serving entities in retailing layer. *IEEE Systems Journal, 16*(1), 649–658.

Rashidizadeh-Kermani, H., Vahedipour-Dahraie, M., Parente, M., Shafie-Khah, M., & Siano, P. (2023). Implementing a flexible penalizing mechanism for wind power producers in the regulating market. *IEEE Access, 11*, 125204–125216.

Rashidizadeh-Kermani, H., Vahedipour-Dahraie, M., Shafie-khah, M., & Catalão, J. P. S. (2019). Stochastic programming model for scheduling demand response aggregators considering uncertain market prices and demands. *International Journal of Electrical Power and Energy Systems, 113*, 528–538.

Rejeb, A., Suhaiza, Z., Rejeb, K., Seuring, S., & Treiblmaier, H. (2022). The internet of things and the circular economy: A systematic literature review and research agenda. *Journal of Cleaner Production, 350*, Article 131439.

Das, S. K., Benkhelifa, F., Sun, Y., Abumarshoud, H., Abbasi, Q. H., Imran, M. A., & Mohjazi, L. (2023). Comprehensive review on ML-based RIS-enhanced IoT systems: Basics, research progress and future challenges. *Computer Networks, 224*, 109581.

Teng, Y., & Lin, B. (2024). New energy supply chain configuration diversified development: The role of the digital economy. *IEEE Transactions on Engineering Management, 71*, 8140–8152.

Thanki, S., & Thakkar, J. (2018). A quantitative framework for lean and green assessment of supply chain performance. *International Journal of Productivity and Performance Management, 67*, 366–400.

Tsao, Y.-C., & Thuy-Linh, V. (2019). Power supply chain network design problem for smart grid considering differential pricing and buy-back policies. *Energy Economics, 81*, 493–502.

Vahedipour-Dahraie, M., Rashidizadeh-Kermani, H., Najafi, H. R., Anvari-Moghaddam, A., & Guerrero, J. M. (2018). Stochastic security and risk-constrained scheduling for an autonomous microgrid with DR and renewable energy resources. *IET Renewable Power Generation, 12*(6), 657–667.

Vahedipour-Dahraie, M., Rashidizadeh-Kermani, H., Najafi, H. R., Anvari-Moghaddam, A., & Guerrero, J. M. (2017). Stochastic security and risk-constrained scheduling for an autonomous microgrid with demand response and renewable energy resources. *IET Renewable Power Generation, 11*(14), 1812–1821.

Westphal, I. (2024). The effects of reducing renewable power intermittency through portfolio diversification. *Renewable and Sustainable Energy Reviews, 197*, Article 114415.

Xu, Z., Hu, Z., Song, Y., & Wang, J. (2017). Risk-averse optimal bidding strategy for demand-side resource aggregators in day-ahead electricity markets under uncertainty. *IEEE Transactions on Smart Grid, 8*(1), 96–105.

Yi, H., Hajiesmaili, M. H., Zhang, Y., Chen, M., & Lin, X. (2018). Impact of the uncertainty of distributed renewable generation on deregulated electricity supply chain. *IEEE Transactions on Smart Grid, 9*(6), 6183–6193.

Yu, Y, & Shahabi, L. (2023). Optimal infrastructure in microgrids with diverse uncertainties based on demand response, renewable energy sources and two-stage parallel optimization algorithm. *Engineering Applications of Artificial Intelligence, 123* (Part B), 106233.

Yunfei, D., Shen, X., Kammen, D. M., Hong, C., Nie, J., Zheng, B., & Yao, S. (2024). A generation and transmission expansion planning model for the electricity market with decarbonization policies. *Advances in Applied Energy, 13*, Article 100162.

Zhang, M., & Guo, M. (2024). Attracting private investment to renewable energy projects in India. *Utilities Policy, 90*, Article 101816.

Zhang, Y., Le, J., Zheng, F., Zhang, Y., & Liu, K. (2019). Two-stage distributionally robust coordinated scheduling for gas-electricity integrated energy system considering wind power uncertainty and reserve capacity configuration. *Renewable Energy, 135*, 122–135.

Zheng, S., Jin, X., Huang, G., & Lai, A. C. K. (2022). Coordination of commercial prosumers with distributed demand-side flexibility in energy sharing and management system. *Energy, 248*, 123634.

Zhu, X., & Cao, Y. (2021). The optimal recovery-fund based strategy for uncertain supply chain disruptions: A risk-averse two-stage stochastic programming approach. *Transportation Research Part E, 152*, Article 102387.

Zornow, F., Talari, S., Ketter, W., Ebrahimi, M., & Shafie-khah, M. (2023). Local flexibility markets and business models. In *Trading in local energy markets and energy communities: concepts, structures and technologies* (pp. 181–220). Springer International Publishing.

Chapter 3
Optimization Techniques in Sustainable Energy Planning

Mostaan Khakpoor

Abstract Sustainable energy planning is essential for enhancing security, addressing climate change, and promoting economic growth. The penetration of renewable energy resources into power systems is crucial for moving toward an eco-friendly grid. In this context, sustainable energy planning stands as a pivotal point. This chapter investigates the optimization techniques in sustainable energy planning, describing the challenges and opportunities that renewable energy resources offer to power systems. In the following chapter, various methods such as linear and nonlinear programming, multi-objective optimization, stochastic techniques, and metaheuristic approaches are reviewed. The chapter also explores the emerging paradigms of novel data-driven and AI-powered algorithms into renewable energy planning purposes. Optimization methods are compared across various facets, and conclusions are drawn. Furthermore, the chapter discusses famous sustainable planning objectives, including environmental, technical, economic, and societal assessments which are essential for the development of a sustainable power system. By presenting the advantages and limitations of different optimization techniques, this chapter equips system planners, decision-makers, and researchers with the knowledge to address the complexities of shifting toward sustainable and renewable energy networks in future power grids.

Keywords Sustainable energy planning · Renewable energy resources · Eco-friendly grids · Mathematical optimization · Multi-objective optimization · Stochastic optimization · Metaheuristic optimization · Data-driven algorithms · Sustainable power systems

---

M. Khakpoor (✉)
Department of Electrical and Computer Engineering, University of Michigan-Dearborn, Dearborn, USA
e-mail: Mostaan.khakpoor@gmail.com

I. Rahimi et al. (eds.), *Sustainable Supply Chain of Renewable Energy Networks*, Energy Systems in Electrical Engineering, https://doi.org/10.1007/978-981-96-4940-2_3

3.1 Introduction

Recently, the United Nations (UN) established a set of objectives for sustainability, referred to as the Sustainable Development Goals (SDGs). The SDGs concentrate on several main goals including ensuring affordable access to modern energy for everyone, increasing the proportion of renewable energy in the total energy mix, enhancing the worldwide energy efficiency, growing economic, and combating climate change (Swain & Karimu, 2020; Tsangas et al., 2022).

Indeed, sustainable energy systems represent the future of our power infrastructure, transforming the energy landscape by integrating renewable energy resources and reducing the reliance on fossil fuels. In the past, conventional energy systems primarily relied on large fossil fuel power plants to meet power demand securely and reliably. The traditional power system structure, known as the centralized model, has been adopted for decades. However, this centralized model has shown several significant deficiencies such as the depletion of fossil fuels, territorial imbalances, and environmental pollution and damage (Chu et al., 2017). Despite recent efforts to reduce the use of fossil fuels, they remain the world's largest energy sources. Figure 3.1 shows the trends of different energy sources over the past years. The data has been collected from Energy Institute—Statistical Review of World Energy (2024). Further investigation shows that CO_2 emissions are also increasing incrementally, see Fig. 3.2. This calls for appropriately addressing the current energy pathway toward sustainability and exploring planning frameworks for the future.

The shifting toward sustainable energy systems is a pressing global priority, aiming to alleviate environmental pollution and promote socio-economic development. Considering the constant depletion of fossil fuels' reserves and increasing climate change issues throughout the world, a shift toward renewable energy sources has become unavoidable. Nonetheless, integrating renewable energy sources including wind turbines and photovoltaic panels into the current power grid presents challenges due to their inherent intermittent nature. Therefore, the development of robust optimization techniques for energy system planning is paramount.

In modern power systems, it is necessary to make the transition toward a socio-economic, environmentally friendly, and sustainable energy system. By increasing the exploitation of renewable energy sources, sustainable energy systems contribute to more balanced systems and reduce fossil fuel-based generations. Furthermore, equipped with Distributed Generations (DGs), sustainable energy systems play a prominent role in shifting toward greener, smarter, and more flexible power systems. In this regard, exploring sustainable energy planning is crucial to address trends, planning frameworks, and optimization techniques (Østergaard & Sperling, 2014).

In general, expansion planning can also be categorized based on static and dynamic viewpoints. In static viewpoint, the planning framework discusses where new equipment should be installed; however, in dynamic viewpoint, the time is also recommended in addition to locations. Considering the planner(s) goals, each viewpoint can be adopted. Additionally, optimization methods can affect the results of final optimal plan for a proposed power system. Indeed, choosing an optimization approach can

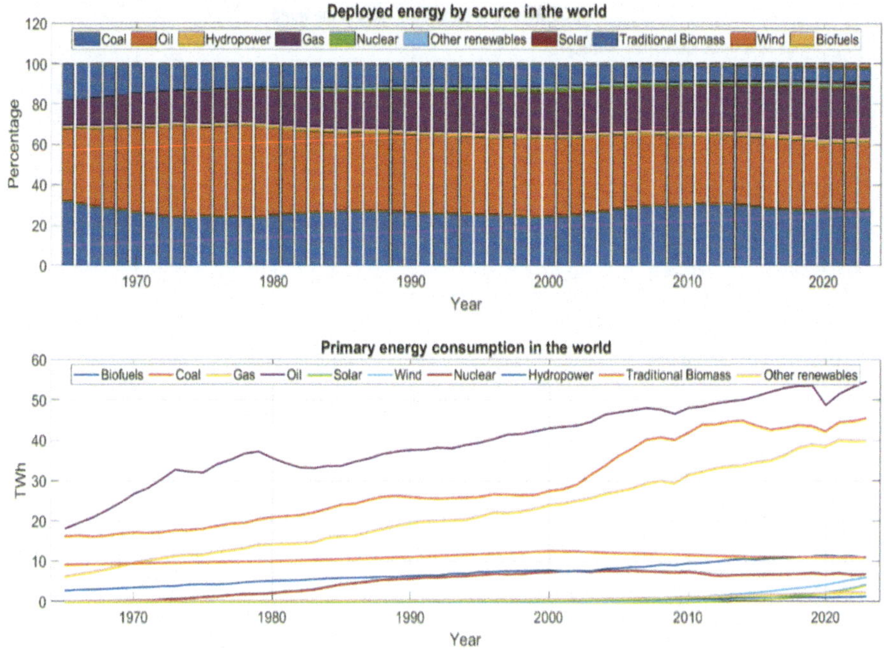

Fig. 3.1 Energy trends in the world, last decade

be influenced by the complexity of the proposed system. In some cases, mathematical optimization may be a better approach for energy planning, while in some other cases, different types of approach may be a better fit. In this way, the optimization techniques can significantly change the final optimal plan which will result in discrepancies in investment, operation, and maintenance of the whole system. Therefore, getting insights into optimization techniques in sustainable energy planning is imperative. To this end, this chapter explores various optimization techniques used by different authors in the literature. Furthermore, it addresses current trends in machine learning algorithms targeting sustainable planning. The main contributions of the chapter are as follows:

- This chapter provides a thorough review of current optimization techniques used in sustainable energy planning. It offers a clear perspective on the state-of-the-art methods and their suitability for various types of renewable energy systems.
- By discussing the implications of different optimization techniques for sustainable energy planning, this chapter serves as a valuable resource for system planners and researchers seeking to optimize power system planning integrated with renewable energy resources. Moreover, it proposes a methodological framework for selecting and implementing optimization techniques based on the needs of renewable energy networks under study.

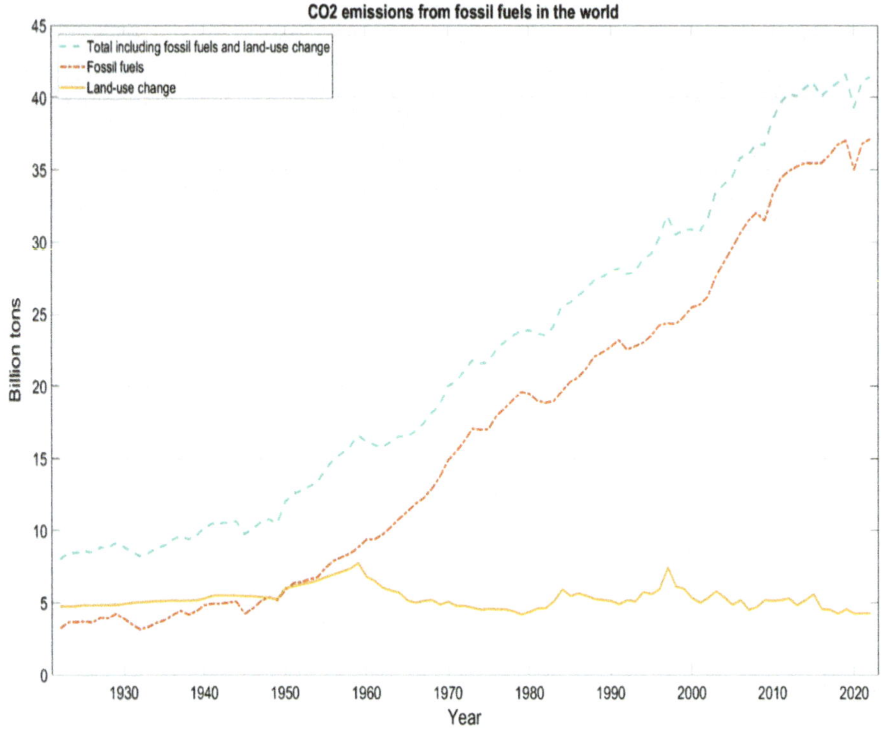

Fig. 3.2 Global CO_2 emissions

- The chapter identifies current gaps and suggests areas for further research. Therefore, it aims to stimulate researchers by showing them avenues of research as potential future works. Furthermore, it contributes to future directions in optimization research for sustainable planning, highlighting the importance of integrating new technologies such as artificial intelligence and machine learning algorithms.

The rest of the chapter is organized as follows: Section 3.2 provides fundamental concepts on power system studies. Section 3.3 addresses some of the main players, contributing to sustainable power systems. Section 3.4 explores the optimization roles in sustainable energy frameworks. A literature review is presented in Sect. 3.5. Section 3.6 presents different types of optimization techniques and discusses the pros and cons of each one. Section 3.7 provides famous planning criteria in sustainable energy planning. Challenges and opportunities are discussed in Sect. 3.8. Section 3.9 presents future avenues of research, aiming to promote sustainable energy development. Finally, the chapter concludes in Sect. 3.10.

3.2 Principles and Basics of Power Systems Studies

Prior to delving into advanced optimization techniques, it is crucial to understand the fundamental concepts that constitute energy system studies. This section explains the conceptual framework of energy planning, defining the role of different energy system studies, time domains, planning types that underlie the planning process.

Generally, power system studies can be categorized based on the time under study. In this context, five groups can be assessed based on the monitoring time span as follows (interested readers may refer to Seifi and Sepasian (2011) for further details):

- **Power System Transients**: This group studies the behavior of a system ranging from nanoseconds to microseconds.
- **Power System Dynamics**: This group refers to studies of a system ranging from milliseconds to seconds.
- **Power System Operation**: If the time horizon monitors the behavior of a system, it falls into this group between minutes to about one week.
- **Power System Maintenance**: This group usually is considered for scheduling maintenance and takes one week to a maximum of one year.
- **Power System Planning**: This group of studies explores the planning for the future of a current power system which ranges from one year to several decades.

Power systems are one of the largest and complex infrastructures made by humans. Operation and protection of power grids require both knowledge and experience. Energy planning should be done so that the future grid can deliver the energy to forecasted demand adequately and reliably. Looking at Fig. 3.3, the general process of power system planning is described.

It should also be noted that power planning may have different purposes, which creates various types. The following types are among the prevalent planning problems:

- Generation Expansion Planning: In this type of planning, the generation sector is the focus. Various power plants shall be candidates for installation within the planning horizon. The selection of candidates between renewable energy sources (e.g., photovoltaic panels, wind turbines) and conventional power plants (e.g., plants powered by coal, gas, oil, etc.) depends on the planning objectives and constraints.
- Transmission Expansion Planning: This type of planning usually investigates the necessary transmission lines to be recommended for installation in the network. The planning may have different goals as well. However, the typical goals are reducing power market congestion and overall system congestion costs.
- Substation Expansion Planning: This type of planning aims to suggest new substations or expand current substations to accommodate the anticipated future demand. Substation expansion planning may also include the optimal sizing of new substation components such as transformers, tap changers, switchgear, and others.

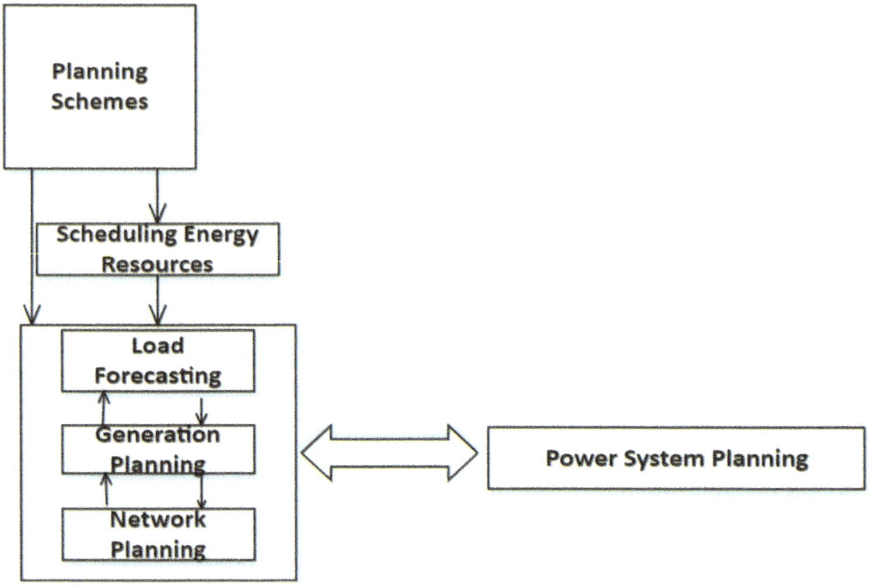

Fig. 3.3 General process of power system planning

- Reactive Power Planning: This kind of planning explores the operation of power systems to maintain voltage levels within acceptable ranges. Devices such as capacitor banks, shunt reactors, and flexible AC transmission systems (FACTS) need to be optimally placed and sized within the network to provide the necessary reactive power support.

In general, each planning problem seeks to investigate four tasks: the optimal location of the installation site, the new optimal capacity, the timing of new installation, and the type of technology that should be considered. It is worth mentioning that a framework incorporating all the aforementioned items is considered comprehensive, while most approaches focus on some of them for simplicity (Hemmati et al., 2013).

3.3 Main Players in Achieving Sustainable Power Systems

A power system is a complex web of interdependent components, including renewable energy resources, energy storage systems, customers, and regulatory frameworks. Understanding how these players can contribute to providing a sustainable system is essential for optimizing the system's performance.

- **Renewable Energy Sources**: To achieve a sustainable power grid, renewable energy sources such as wind farms and solar parks play a prominent role in accelerating the sustainability transformation. The nature-associated energy sources are ubiquitous and in different forms including wind, solar, biomass, ocean-waves and hydro. Renewable energy sources while replacing them with conventional power plants can significantly contribute to increasing sustainability capacities. Powered by nature, these sources help alleviate concerns about energy depletion. Therefore, renewable energy sources are crucial players in developing sustainable grids.

- **Demand-Side Management**: The end-users also play a crucial role in sustainable energy systems. Demand-side management generally refers to a list of programs that can change the behavior of end-users either directly or indirectly. Demand response (DR) program is a special subset of demand-side management. By the definition stated by U.S. Department of Energy (DOE), DR is the "changes in electric usage by end-use customers from their normal consumption patterns in response to changes in the price of electricity overtime, or to incentive payments designed to induce lower electricity use at times of high wholesale market prices or when system reliability is jeopardized" (DOE, 2006). Participation of end-users in DR programs can significantly help in developing sustainable power grids.
- **Energy Storage System**: When dealing with uncertainties in renewable energy outputs, energy storage systems can be an effective tool to capture the surplus energy and deliver it when the system is in need. Fluctuations in energy production by renewable energy sources can be mitigated by installing and developing energy storage systems. There are several forms of energy storage systems such as battery storage (e.g., lithium-ion, sodium-sulfur batteries), thermal energy storage (e.g., molten salt), mechanical storage (e.g., pumped hydro, flywheel energy storage), hydrogen storage (e.g., hydrogen fuel cells), and super-capacitors. Sustainable electricity systems require active participation and development of energy storage systems.
- **Regulatory Frameworks**: Regulations influence current and future road maps of the power grids. In fact, power grids have undergone significant changes since their initial implementation, mostly due to new regulations set by policymakers and practitioners. In developing sustainable energy systems, regulatory frameworks are indispensable. By setting targets for emissions renewable energy penetration and participation in total energy production, subsidies, and incentive programs, putting deadline for obsolete technologies usage, regulations create a structured landscape that planning must navigate.

3.4 Optimization Roles in Sustainable Energy Planning

Optimization techniques play an important role in sustainable energy planning by providing a structured paradigm to decision-making and offering optimal planning solution(s). These techniques help in identifying the most effective ways to integrate renewable energy sources, considering various constraints and objectives (Bishop et al., 2008, 2010). Optimization can address numerous facets of energy planning, including location, capacity planning, energy storage integration, and grid management.

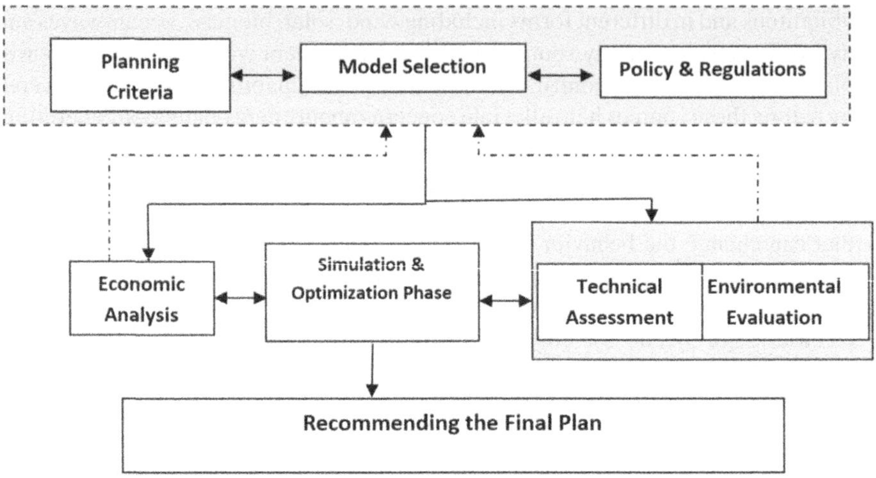

Fig. 3.4 Optimization procedure in sustainable energy planning

Nonetheless, choosing an optimization algorithm depends on various factors including the system under study, the planning horizon, participant components in the planning, etc. The flowchart of the optimization procedure in sustainable energy planning is depicted in Fig. 3.4.

In the optimization procedure, planning criteria are first proposed with respect to the planner's preferences. Then, optimization at the core of the problem assesses the planning goals based on various considerations. Through interaction between the assessment results and modification of the plans, the final optimal plan will be recommended. Note that this is the general procedure for optimization in the planning process. Hence, different algorithms may have slightly different flowcharts. However, the general procedure remains the same for all optimization techniques. Later in the chapter, we will also shed light on some of the famous planning criteria in sustainable energy planning.

3.5 Literature Review

Moving toward greener and more efficient sustainable energy systems requires investigation and research. Bali Swain et al. (Swain & Karimu, 2020) proposed a two-step estimation method to capture the effects of renewable electric prices and different SDGs. The results further recommend that combining different SDGs can be more impactful in achieving their targets and goals.

Giatrakos et al. (2009) proposed a sustainable energy plan for an island in Greece. The paper considered three scenarios for developing renewable energy portion in total power generation. HOMER has been used as the environment for modeling and simulation purposes. They assumed that renewable energy production would range from 20% up to 100%, to be employed on the island, replacing the conventional power plants. In their model, they considered energy production by photovoltaic parks, wind turbines, and hydrogen generators. To further motivate the deployment of renewable energy resources, the authors set a minimum electricity production by renewable energy resources and imposed CO_2 emission fee.

Ervural et al. (2018) proposed an investment sustainable energy plan by using a weighted goal programming approach. A fuzzy Technique for Order Performance by Similarity to Ideal Solution (TOPSIS) method evaluates social acceptance while considering renewable energy potential in the problem. The Analytic Hierarchy Process (AHP) has been developed to assign weights to each criterion. Environmental criteria have been considered as the planning goals. Fuzzy TOPSIS is chosen as the final optimal alternative.

Sustainable energy planning may also consider the requirements of renewable energy resources for implementation such as site size for wind farms or solar parks. Stoeglehner et al. (2011) postulated an integrated spatial and sustainable energy planning model. They considered four spatial-resource-planning continuums such as urban centers, suburban areas, small towns, and rural areas. Taking spatial dimensions into consideration can result in optimal power exploitation by connecting different grids.

Beccali et al. (2003) conducted an energy planning study in Sardinia utilizing the Elimination and Choice Translating Reality (ELECTRE) method. The research encompasses three scenarios focusing on environmental protection, economic considerations, and energy concerns, while evaluating fourteen distinct renewable energy sources or alternatives. Chang (2015) proposed a goal programming model to address the capacity expansion challenges of renewable energy. This model took into account various plant types, site selection, and additional criteria.

Klemm and Wiese (2022) have recently introduced energy sustainability indicators, including absolute greenhouse gas emissions, absolute energy costs, and absolute energy demand, for use in optimization models of urban energy systems. Tsangas et al. (2023) proposed unexpected events that can change the sustainable energy planning frameworks and targets. They considered a lack of fossil fuels' suppliers and global pandemics, e.g., COVID-19, as the unpredictable, yet effective, situations affecting the sustainability progress of power grids.

Aryanpur et al. (2019) examined the electricity supply sector in Iran via the AHP-TOPSIS method for energy planning. An optimization model, known as MESSAGE, was applied to assess the transition toward a sustainable system. The criteria were categorized into techno-economic, social, and environmental aspects. These criteria were further assessed in the final comparison of the evaluated pathways, taking into account sensitivity to fuel price changes, acidification potential, human toxicity, and nuclear waste.

Demirtas (2013) used an AHP methodology to deal with multi-criteria sustainable energy planning. The paper attempted to find the best renewable energy source for sustainable energy planning. The results indicated wind energy, biomass, and geothermal as the top three suitable renewable energy for sustainable energy planning.

Ma et al. (2013) suggested a flexibility index for sustainable energy planning in power systems. They considered part-loaded synchronized and quick start/shut down generators to deal with the uncertainties posed by the development of renewable energy sources in the system. However, the objective of planning was to choose the traditional minimization function of operation and investment costs, while taking the flexibility index into consideration.

Szabo et al. (2013) investigated the least-cost sustainable energy planning in Sub-Saharan Africa. The paper attempted to increase the utilization of the untapped renewable energy potential to meet Africa's energy demand in an environmentally friendly fashion. The authors considered the electricity production cost for diesel generators and PV-based panels and small hydropower plants. The goal is to reach a sustainable energy scheme that embodies all three mentioned technologies while minimizing the final electricity cost in the system.

Zadehbagheri et al. (2023) investigated sustainable operation scheduling in an energy hub. They considered photovoltaic systems, tidal stream turbines, energy storage systems, electrolyzers, and H2 tank storage in their hub modeling. The paper specifically aims to model the effects of demand response programs on operation scheduling. The total operation cost and pollution cost were considered within a mixed integer linear framework. The epsilon constraints method, augmented by the fuzzy satisfying method, was used to address the multi-objective nature of the problem.

Potrč et al. (2021) developed a multi-period mixed integer linear program to model biofuel production and demand, along with a renewable electricity supply network. Critical comments on renewable-based fuel and raw material availability were taken into consideration. The objective of the paper is to maximize net present values of sustainability, including economic, environmental, and social considerations.

Chien et al. (2021) investigated sustainable energy solutions for severe energy crises in the Pakistani power system. They analyzed the potential of wind, solar, biomass, and geothermal energy in the Pakistani network. They developed a Fuzzy-AHP and non-parametric DEA method to explore replacing fuel-based energy with green hydrogen, produced through renewable energy sources.

Alizadeh et al. (2020) examined the shift from conventional power systems to renewable-dominated power grids, particularly for countries economically dependent on fossil fuels, like Iran. They developed a multi-criteria framework that included economic, security, human well-being, and energy vulnerability factors. They concluded that solar panels offer the best potential for renewable energy penetration in Iran.

Manjong et al. (2021) advanced sustainable energy planning for the Cameroonian power system as a representative case of Central Africa. In their study, they proposed achieving 100% renewable energy production by 2050. The results indicated that solar power parks have the greatest potential for the region. Different scenarios, including carbon costs and various policies, were evaluated. The authors also examined how renewable energy could affect the heat and transportation sectors in Cameroon.

3.6 Optimization Techniques

Sustainable energy planning is inherently interdisciplinary, blending technological, environmental, economic, social, and political dimensions to design, implement, and manage energy systems that minimize negative impacts while maximizing efficiency and societal benefits. Optimization methods play a crucial role in this process, offering mathematical frameworks to identify the most effective solutions within complex and multifaceted constraints. However, the true challenge and opportunity lie in integrating these optimization techniques across diverse fields such as economics, social science, and policy.

The world is at a crossroads with energy policy and production. As we confront the dual challenges of ensuring reliable energy supply and mitigating environmental impacts, optimization in sustainable energy planning emerges as a key to unlocking a greener future. This section serves to demystify the various techniques employed to seamlessly integrate renewable resources into existing power systems while maintaining, or even improving, grid reliability and economic efficiency.

The optimization techniques in sustainable energy planning differ from each other in nature and the requirements for implementation phase. The choice of one optimization algorithm may depend on the model under study, planning criteria, involving or ignoring uncertainties, equations and formulations of the problem, data availability, etc. Nonetheless, based on the given problem, researchers may prefer to utilize one or more kinds of optimization methodologies. Below, we explore some of the most prevalent approaches used by authors in the sustainable planning area.

3.6.1 Mathematical Methods

Mathematical methods are among the first algorithms used for sustainable research and development. These methods are based on the model under study. Therefore, they heavily rely on the physics laws for implementation. Mathematical optimization methods are identified as white-box optimizers where the model is thoroughly understood.

Linear programming (LP) is often used in sustainable planning where relationships are linear (Rizzo & Savino, 2012; Xydis & Koroneos, 2012), for instance, in fuel-mix optimization or unit commitment problems. LP models are relatively straightforward and can be solved efficiently with current software.

Nonlinear programming (NLP) becomes necessary when dealing with nonlinear cost functions, such as some generators' cost function or when modeling the physics of power flows (Al-Hamouz & Al-Faraj, 2003; Alhumaid et al., 2021). NLP problems are more complex and require advanced algorithms to solve.

Some of the other popular methods that have been used in sustainable energy planning include mixed integer programming (Cosic et al., 2021), mixed integer nonlinear programming (Huang et al., 2019), multivariate scenario-tree methodology (Lei et al., 2021), Bender's decomposition (Hemmati et al., 2018), game theory (Liu et al., 2021; Wang et al., 2022), dynamic programming (Groumpos et al., 1996), and Bayesian game modeling (Huang et al., 2024).

3.6.2 Uncertainty-Aware Approaches

Power systems today deal with uncertainties more than ever before. There are many uncertainties that have been included in the systems. From the past, electricity systems had uncertainties related to the total amount of demand, the outage or failure of a line or power plant in the system, voltage drop and frequency oscillation, natural disasters, etc. However, with the advent of new technologies and new participants in the distribution system such as renewable energy resources, price-driven markets, incentive-based demand response, and prosumers, the new grid is even more uncertain. Here are some of the most uncertainty-aware methods used by authors in sustainable energy planning.

Stochastic Programming: Stochastic programming is one of the most widely accepted approaches for modeling and managing uncertainties in sustainable energy planning (Yu et al., 2019). In this approach, uncertainties are modeled based on the core principle of scenario generation. However, additional steps may be necessary when modeling uncertainties using scenarios. To represent uncertain variables, a scenario-based approach is used, which incorporates scenario generation (e.g., produced from probability density functions) and scenario reduction techniques (e.g., backward reduction, fast forward approach). In stochastic programming, uncertainty in renewable energy generation is represented by a set of discrete scenarios, each with an associated probability. For example, different scenarios might depict varying levels of solar irradiation or wind speed over a planning horizon. The optimization model aims to find a solution that minimizes costs or maximizes reliability across all these scenarios. This approach enables better planning under uncertainty by accounting for multiple possible outcomes, hedging against unfavorable conditions, and thereby enhancing grid reliability and cost-effectiveness. Stochastic programming is widely used in energy planning (Mavromatidis et al., 2018; Vera et al., 2022).

Monte Carlo Simulation: Monte Carlo Simulation (MCS) is a computational technique that produces random sampling to obtain numerical results. MCS relies on generating random samples from a probability distribution to represent uncertain variables, such as wind farm outputs for the day ahead or risk analysis in renewable energy infrastructures (Arnold & Yildiz, 2015). MCS works by generating many random scenarios (usually predetermined by the planner) for solar and wind generation based on historical data and probability distribution functions. Each scenario simulates a possible state of the grid, capturing the variability in renewable outputs. For example, MCS can be used to estimate the distribution of total energy produced by a wind power plant over a month or a year. By simulating hundreds or thousands of scenarios, planners can identify the likelihood of energy production over the period of planning. MCS is widely used in sustainable energy planning (Baležentis & Streimikiene, 2017; Constante-Flores & Illindala, 2018; Pask et al., 2017).

Robust Optimization: Robust optimization (RO) is another technique. RO aims to find solutions that perform well across the worst-case scenario of uncertainty (Ratanakuakangwan & Morita, 2021). This is particularly relevant for renewable energy systems, where ensuring supply under worst-case scenarios (e.g., prolonged periods of low wind or solar output) is crucial. Unlike stochastic programming, which considers the expected performance across multiple scenarios, RO aims to protect against the worst-case scenario. It works by finding solutions that are feasible and perform reliably regardless of how uncertainties unfold within specified bounds. For example, in a wind power plant, RO might ensure that, even during periods of unexpectedly low wind speeds, the power system can still meet demand without incurring excessive costs. RO is useful when planners want to design a system that can tolerate worst-case future scenarios. For instance, RO can define "uncertainty sets" (e.g., a range of possible blackouts or demand surges) to cover a broad spectrum of extreme events. In this way, RO ensures that the power grid performs well even in the blackouts or demand surges, though it may sacrifice some optimality under normal conditions. RO is also used by researchers in the literature (Dong et al., 2016; Hosseini et al., 2020).

There are also other methods like the mathematical-statistical approach (Otić et al., 2021) and the interior point method (Pages et al., 2009), dealing with uncertainties in the literature.

3.6.3 Metaheuristic Approaches

Metaheuristic methods are capable of providing near-global optimum solutions for complicated optimization needs in sustainable energy planning. They have been designed to work in a variety of optimization problems, flexible to changes, and easily adaptable. These kinds of approaches are generally powerful in dealing with complex and large-scale problems in large and sophisticated infrastructures. The execution time is usually higher than that of mathematical methods. They generally work based on population-based solutions or a single solution.

Genetic algorithms (GAs) are inspired by Darwin's theory so that an initial population improves in each round by mutation and crossover operators. GA is one of the most popular metaheuristic algorithms in the sustainable energy planning problems (Gómez et al., 2023; Lazzari et al., 2023; Yanık et al., 2016). The general optimization process of GA algorithm in a dynamic planning framework is shown in Fig. 3.5. This diagram is also applicable for static viewpoint, except the nested loop for the "year" which is simplified to one iteration.

Fig. 3.5 GA optimization diagram in dynamic planning problems

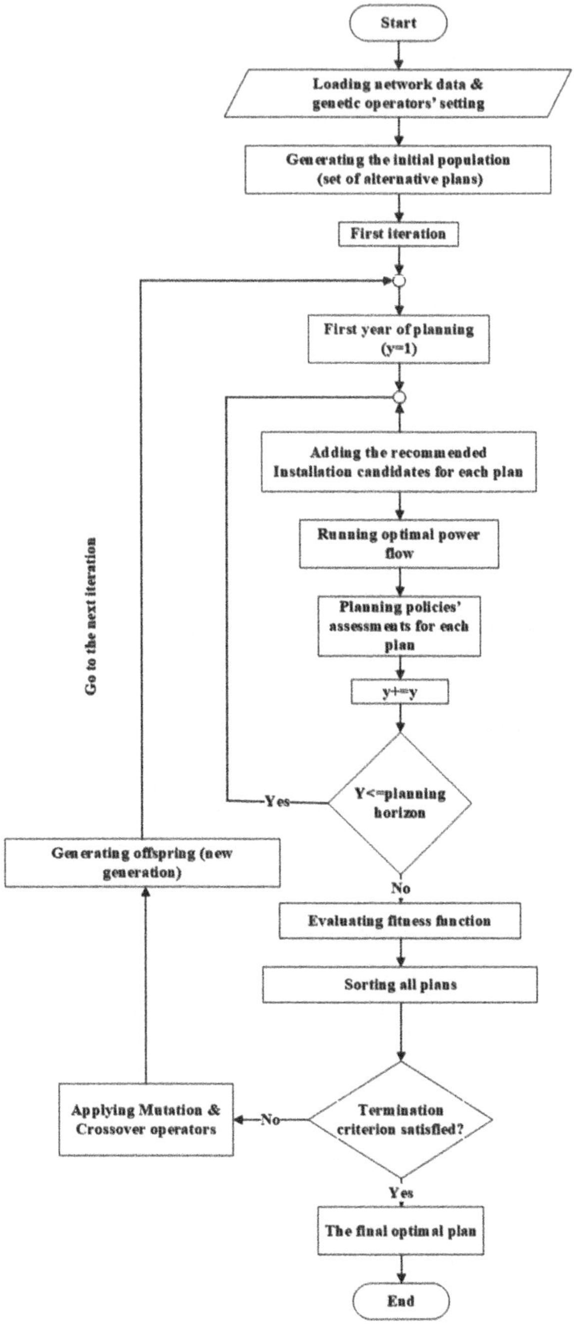

Simulated annealing (SA) is another metaheuristic algorithm, inspired by the physical process of heating and then slowly cooling a material in metallurgy. Unlike GA, SA starts with only one initial solution. SA is widely used in sustainable planning due to its simplicity and ability to escape local optima, making it a robust choice for this complex problem (Katsigiannis et al., 2012; Zhang et al., 2018).

Other popular metaheuristic optimization methods used by authors include particle swarm optimization (PSO) (Abdullah et al., 2018; Qin et al., 2019), gravitational search algorithm (Marzband et al., 2014), normal boundary intersection method (Gitizadeh et al., 2013), θ-particle swarm optimization (Baziar & Kavousi-Fard, 2013), and pattern search algorithm (Jayasekara et al., 2014).

3.6.4 Multi-Objective Optimization

The Multi-Criteria Decision-Making (MCDM) approaches to tackle decision-making while considering multiple goals. In this type, a decision should be made considering both quantitative and qualitative objectives. In MCDM problems, objectives are usually in conflict with each other so that increasing one goal preference decreases the other goals. MCDM algorithms have been applied to various areas where such decision-making is required. MCDM methods can be further classified as multi-attribute decision-making and multi-objective decision-making. Interested readers may refer to Clímaco (2012) for further details.

In sustainable energy planning, various criteria may be taken into consideration such as economic indices, environmental indices, and more. Therefore, dealing with sustainable energy planning optimization can be seamlessly accommodated in an MCDM framework. In this light, determining the optimal plan(s) in sustainable energy planning can be viewed in terms of MCDM. Finding trade-offs between competing objectives can be done by superior-posterior techniques. Zadeh (1965) was one of pioneers who contributed to decision-making on fuzzy set, close to human judgment. Later, his idea was included in many decision-making problems.

The Analytic Hierarchy Process (AHP) is a structured technique for dealing with complex decision-making and is used for organizing and analyzing decisions based on mathematics and psychology. Developed by Saaty (1980), AHP helps in making decisions that are consistent and can be rationally justified. Figure 3.6 depicts the general process of AHP. The AHP procedure can be briefly summarized as: state the problem, broaden the objectives, identify the criteria, structure the problem (criteria, sub-criteria, and alternatives), compare each element, perform calculations, and make the decision. AHP is one of most popular methods used in the literature (Abdullah & Najib, 2016; Al-Qudaimi & Kumar, 2018; Budak et al., 2019).

The Technique for Order Preference by Similarity to Ideal Solution (TOPSIS) is another multi-objective optimization technique, introduced by Hwang and Yoon (1981). The primary principle behind TOPSIS is that the chosen solution should have the shortest geometric distance from the positive-ideal solution and the longest

Fig. 3.6 Typical diagram of AHP in a planning problem

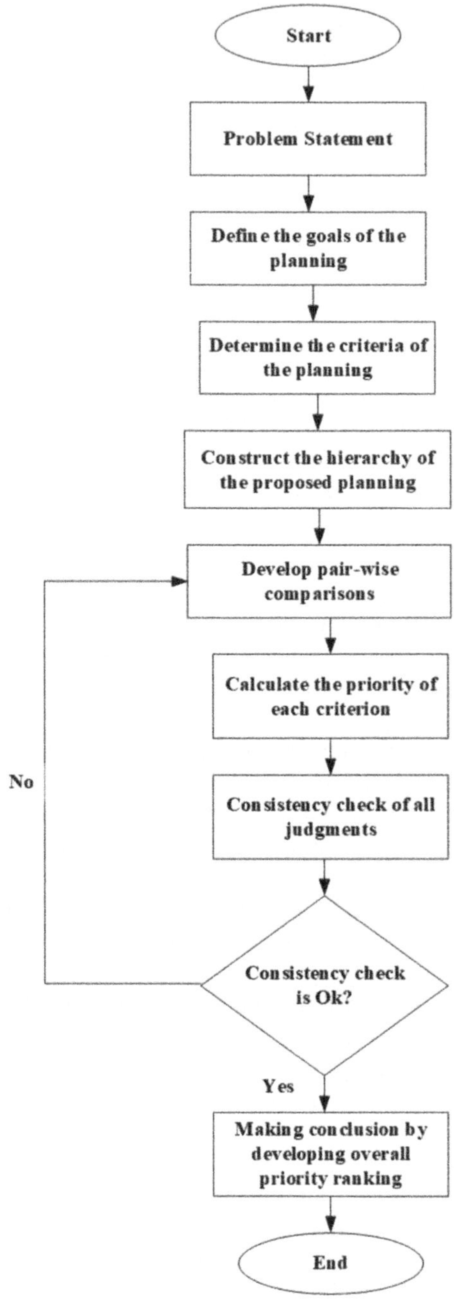

geometric distance from the negative-ideal solution. Therefore, to compare the alternatives and refine the final ranking, the Euclidean distances between each alternative and both the ideal and negative-ideal solutions are calculated. Then, the closeness coefficient is employed to calculate these distances, respectively. Due to its clear and logical methodology, TOPSIS has been widely used in sustainable energy planning problem (Afsordegan et al., 2016; Fathipour & Saidi-Mehrabad, 2018).

The combined AHP-TOPSIS approach is also applied by researchers, decreasing the uncertainties in group decision-making and ensuring more robust solutions (Solangi et al., 2019).

Other multi-objective techniques used by researchers in sustainable energy planning are ELECTRE (Ezbakhe & Pérez-Foguet, 2021), Preference Ranking Organization Method for Enrichment Evaluation (PROMETHEE) (Montajabiha, 2016), Non-Sorting Genetic Algorithm II (Khakpoor et al., 2017a), Fuzzy MCDM (Turgut & Tolga, 2018), and Multi-Attribute Utility Theory (MAUT) (Hahn, 2015).

3.6.5 *Data-Driven Methods*

Artificial intelligence (AI) and machine learning (ML) algorithms, as a subsidiary of the AI, have been applied in various optimization problems. Indeed, promising solutions resulting from using ML algorithms extend their application in energy planning such as sustainable energy planning. Equipped with real-time data and the availability of historical data stored in the system, data-driven methods are viable approaches, improving forecasting accuracy in renewable energy outputs and achieving optimal planning schemes in a sustainable energy environment.

Artificial Neural Network (ANN) is one of the first ML algorithms used by researchers. ANN is a subset of machine learning and artificial intelligence that is designed to recognize patterns and interpret data in a way that mimics the human brain. ANN can predict the outputs by looking at their associated input using hidden layers of neurons that give each input weight to map it to its appropriate output. ANN has shown its capability and flexibility to work with problems in sustainable energy planning (Debnath & Mourshed, 2018; Kalogirou, 2000).

Reinforcement Learning (RL) is another type of machine learning algorithm where an agent learns to optimize decision-making in real-time data by taking actions in an environment to maximize cumulative rewards. It involves learning through interaction with a dynamic environment. Therefore, RL can also adapt to changes and accordingly optimize operations autonomously. RL is another of the most commonly used optimization algorithms in sustainable energy planning (Nakabi & Toivanen, 2021; Perera & Kamalaruban, 2021).

Other famous methods of ML algorithms have also been employed in the literature such as support vector machine (SVM) (Bokhoree & Balnac, 2009), multi-agent-based algorithms (Jun et al., 2011; Paliwal et al., 2022), decision trees (Huo et al., 2021), deep neural networks (Wen et al., 2023), harmonic search algorithms (Rastgou et al., 2018), and graph-based methods (Aviso et al., 2017; Lim et al., 2020).

3.6.6 Hybrid Approaches

In some planning problems, for various reasons (e.g., increasing accuracy, parameter tuning), different methods may be combined to build a hybrid approach. The combination of multiple optimization techniques has been successfully used to tackle complex planning problems. In fact, this category is not a standalone but a derivative version of the previously mentioned ones. Hybrid approaches are typically used to improve the performance of the original algorithm or to add capabilities absent in the original method.

Based on the aforementioned optimization methods, it can be deduced that all previous categories have their advantages and contributions to sustainable energy planning. Hybrid approaches are developed to strengthen effectiveness and fine-tune an existing optimization method. For example, genetic algorithms are commonly used in single-objective optimization planning. In Khakpoor et al. (2017b), a fuzzy-based law was included in the optimization formulations to enable the genetic algorithm to handle multiple objectives by assigning a fuzzy number between [0,1] to each planning criterion. By doing this, a new fuzzy-genetic algorithm was developed, making it applicable in a multi-objective framework. Most hybrid approaches are derivatives of metaheuristic algorithms like GA and PSO (Babatunde et al., 2020). Some hybrid approaches used in sustainable energy planning include iterative-GA (Khatib et al., 2012), ANN-GA-MCS (Lujano-Rojas et al., 2013), harmony search and PSO (Alshammari & Asumadu, 2020), and ARIMA-ANN (Fara et al., 2021).

3.6.7 Discussion on the Optimization Methods

Each optimization method has its own advantages and disadvantages. Indeed, selecting an optimization type depends on various factors, such as planner preferences, model data, type of uncertainties, planning viewpoint, planning criteria, time horizon, and so on. For specific modeling assumptions, several methods may be investigated to end up with the best results. This sometimes requires employing different approaches to fit the problem for achieving the desired outputs. Therefore, there is no panacea when it comes to optimization algorithms. One method may outperform others in one problem, while in another problem with different planning considerations and settings, the defeated algorithms may have better results. In Table 3.1, we compare different optimization approaches from various perspectives.

Looking at Table 3.1, several insights can be interpreted as follows:

- Simplicity in Implementation: Mathematical methods can be complex but are well-documented. Metaheuristic and data-driven methods can vary widely in complexity.
- Execution Time: Mathematical methods generally have efficient algorithms, while uncertainty-aware methods and metaheuristics might require more computation time.

Table 3.1 Comparison of optimization techniques

Features	Mathematical methods	Uncertainties-aware methods	Metaheuristic algorithms	Multi-objective methods	Data-driven methods
Simplicity in implementation	Moderate to high	Moderate	Low to moderate	Moderate to high	Low to moderate
Execution time	Fast	High	Moderate to high	Moderate	Moderate
Computational complexity	Moderate	High	Moderate to high	High	Moderate to high
Solution quality	Moderate to high	Moderate to high	Low to moderate	Moderate to high	Moderate
Scalability	Low to moderate	Moderate	Low to moderate	Moderate	Moderate to high
Renewable-modeling appropriability	Moderate to high	High	Moderate to high	Moderate to high	Moderate to high
Chance of local optimum	Low to moderate	Low to moderate	High	Low to moderate	Moderate
Comprehensive planning capability	Low to moderate	Moderate	Moderate	High	Moderate
Interpretability	High	Moderate	Low to moderate	High	Low to moderate
Bug fixing	Moderate	Moderate	Low to moderate	Moderate	Low to moderate
Hardware requirement	Low to moderate	Moderate to high	Moderate to high	Low to moderate	High
Software availability	High	Moderate	Moderate	Moderate	High
Free access	High	Moderate to high	High	High	High
Convergency	High	Moderate	Moderate	Moderate to high	Moderate

- Computational Complexity: Mathematical methods are usually computationally efficient for linear problems but can become exponential for nonlinear problems. In other methods, complexity ranges from moderate to high, increasing with the number of variables, scenarios, population size, and large datasets.
- Solution Quality: Multi-objective methods are capable of finding balanced solutions for conflicting objectives. Mathematical methods work well for convex problems but perform poorly with nonlinear ones. Uncertainty-aware methods depend on the accuracy of probabilistic models.
- Scalability: Data-driven methods tend to have better scalability compared to others. Parallel computing can assist them in handling substantial resources.
- Renewable-Modeling Appropriability: Techniques are judged on how well they adapt to new or evolving models, with high appropriability meaning they adapt well. Uncertainty-aware approaches align well with this feature.
- Chance of Local Optimum: Metaheuristics often get trapped in local optima, while mathematical methods can avoid this with global optimization techniques.
- Comprehensive Planning Capability: Multi-objective methods are specifically designed for this, while other methods might require adaptations.
- Interpretability: Mathematical methods and multi-objective methods often provide more understandable results compared to metaheuristic, uncertainties-aware, and Data-Driven methods.
- Bug Fixing: Simpler methods tend to be easier to debug. Data-driven methods can be challenging due to their complexity.
- Hardware Requirement: Data-driven methods typically require more computational resources, especially with large datasets and complex models.
- Software Availability: All categories have numerous tools available, though some specific methods within each category might have more limited options.
- Free Access: Many mathematical solvers, metaheuristic algorithms, and multi-objective tools are freely available. Additionally, numerous open-source tools and libraries are available for data-driven methods, however, they vary based on the specific approach and tool.
- Convergency: Mathematical and multi-objective methods often have clear convergence criteria, while metaheuristics and data-driven methods might not always guarantee convergence to the global optimum.

3.6.8 Optimization Use Case Insight

A more nuanced understanding of the use case related to each optimization method helps in selecting the most appropriate approach based on specific problem characteristics, computational resources, and desired outcomes. Each optimization group can be further described as follows:

- Mathematical Methods: These are generally a good fit for small to medium-sized, well-defined optimization problems where interpretability is important. However, they are not usually suitable for large-scale or highly complex systems.

- Uncertainty-Aware Methods: These are typically best suited for scenarios where uncertainty implementation plays a significant role, such as in sustainable power systems with high penetration of renewable energy resources. Limited scalability can be a barrier for very large systems.
- Metaheuristic Algorithms: These are usually useful for complex and real-world optimization problems where traditional algorithms may fail. They are suitable for scheduling and resource allocation, but scalability and convergence issues limit their use in very large problems.
- Multi-objective Methods: These are the best choice for problems with multiple objectives simultaneously, often conflicting, such as environmental and economic trade-offs in sustainable energy planning. However, the execution time for very large and high-dimensional multi-objective problems can be very high.
- Data-Driven Methods: These are well-suited for problems with large amounts of historical data available, such as forecasting and predictive wind farms' generation or solar parks' outputs in sustainable energy planning. Data availability can be a bottleneck for very large or complex problems.

Therefore, each optimization method can be used in different arrangements. Mathematical methods are ideal for small to medium-sized, well-structured problems. Uncertainty-aware methods excel in scenarios with significant uncertainty but can be computationally expensive. Additionally, metaheuristic algorithms are useful for complex and non-convex problems but struggle with scalability. Multi-objective methods are tailored for problems with conflicting objectives but can become computationally expensive with many objectives. Finally, data-driven methods can be powerful for predicting power output from wind power plants or solar panels, provided that sufficient and high-quality data is available. It is worth mentioning to remind that selecting an optimization technique for a sustainable energy planning problem requires complete knowledge of data availability, the scale of the problem, types of uncertainties, formulations, equations, and more. After gaining this understanding, a researcher may choose an optimization method to achieve the best possible result.

3.7 Planning Criteria

Sustainable energy planning is critical for ensuring that energy systems are technically feasible, environmentally friendly, economically viable, and socially equitable. Accordingly, conducting research in sustainable energy planning can be categorized into four main categories: technical, economic, environmental, and social considerations. Planning criteria in sustainable energy planning are tabulated in Table 3.2.

It should be noted that comprehensive sustainable energy planning may consider all categories with appropriate indices in their planning optimization framework. However, considering planning criteria for each problem is a specific task that

Table 3.2 Planning criteria in sustainable energy planning

Assessment type	Planning criteria
Technical assessment	Primary energy ratio, reliability, security, technology maturity, efficiency, safety, electricity interconnection capacity
Economic assessment	Investment cost, fuel cost, operation cost, maintenance cost, payback period, equivalent annual cost, service life, fuel price changes
Environmental assessment	CO_2, NO_x, SO_2 emissions, noise, ecosystem impact, land use, particle emission, waste disposal
Social assessment	Job creation, social benefits and acceptability, poverty status, energy affordability, political acceptance

depends on various facets such as planner preferences, policies and regulations, projects' budget limit, geography of the model under study, and other considerations. For example, sustainable energy planning in regions with underdeveloped infrastructure requires not only technical and economic considerations but also a nuanced understanding of political, regulatory, and societal facets. Optimization techniques should capture these complexities in a manner that balances competing interests and constraints. To address this, multi-objective optimization techniques may be used to incorporate policy indices and regulatory constraints into the decision-making process. By incorporating different assessments in the optimization problem, the outcome will be suitable for the network. Stochastic optimization is also useful in scenarios where policies may change during the planning horizon. By modeling regulatory uncertainty as a probabilistic input, planners can develop robust energy systems that remain optimal across a range of potential policy outcomes. Moreover, constraints can be applied to the problem, ensuring compliance with region-specific energy laws, subsidy structures, or emissions. By accounting for political, regulatory, and societal dimensions, these techniques enable the development of robust, socially responsible, and sustainable energy solutions.

3.8 Challenges and Opportunities in Sustainable Energy Planning

Renewable energy sources are one of the cores of sustainable energy development and research (Bazmi & Zahedi, 2011). The integration of renewable energy resources into power systems also poses some challenges as follows:

- Intermittency and Variability: Renewable energy sources such as wind turbines and photovoltaic panels are dependent on weather conditions. Therefore, they are intermittent and variable according to their surrounding weather. This makes maintaining the power balance, i.e., consistently matching supply with demand, a more complex task.

- Grid Stability: High penetration of renewable energies can impact grid stability and reliability, requiring advanced control and management strategies and employing energy storage systems.
- Economic Viability: The initial investment costs for renewable energy infrastructure can be high, posing financial challenges. However, these investments typically pay off due to lower operation costs over the years.
- Regulation Bottlenecks: While many developed countries set programs for moving toward sustainable energy systems, there is still an increase in emissions and oil usage worldwide. This indicates that unanimous agreement and international cooperation on relevant targets and policies are necessary to increase public awareness toward a greener grid.
- Interdisciplinary Nature: Sustainable energy planning is inherently interdisciplinary, requiring consideration of various aspects. Incorporating policy constraints into optimization models at local, national, and international levels, societal interventions, and technical constraints, such as emission limits and grid integration standards, can all have significant impacts. Large-scale renewable energy projects face not only technical hurdles but also political, societal, and regulatory constraints. A tangible example is wind farm siting, where optimization models may account for the technical feasibility of wind exploitation. However, factors like proximity to residential areas, aesthetic concerns, and noise levels carry significant weight during the implementation phase. Therefore, optimization can guide project developers in selecting locations that balance different considerations. Societal acceptance is crucial for ensuring smoother project approval and reducing community opposition. Thus, incorporating these constraints into optimization frameworks is essential for realistic and feasible energy planning.

These challenges make the fast transition toward sustainable grids more complex and difficult. However, sustainable energy systems have several benefits which make them suitable for adoption and implementation within our grids. Some of the opportunities within the sustainable energy systems are as follows:

- Environmental Benefits: The movement to a greener and eco-friendly grid is crucial for future societies. Reducing greenhouse gas emissions, produced by fossil fuels generators, helps improve air quality. Additionally, lower reliance on fossil fuels contributes to mitigating global climate change.
- Energy Independence: When local renewable resources are used, energy security is enhanced since the network is no longer dependent on the main or neighboring grids. Additionally, with locally sourced energy, fuel transportation costs are significantly reduced. Therefore, it also reduces dependence on imported fuels.
- Technological Advancements: Renewable energy resources embedded with power electronic devices provide better transition to digitalization and smart grids. Innovations in energy storage, grid management, and predictive analytics can enhance the integration of renewables.
- Security Benefits: Energy production is vital for each country. Delivering secure and consistent electricity to feed the various demand types is of paramount

importance. Importing and transferring fossil fuels may endanger due to international policies and price inflation. Sustainable energy resources mitigate these risks through lower transmission costs and lack of importing issues. Therefore, a sustainable system is more secure than a traditional system.

It can be deduced that the current hurdles involve integrating intermittent renewable energies, crafting appropriate regulations, and ensuring social acceptance. On the other hand, opportunities lie in technological advances, decreasing costs for renewables, and increasing public awareness of climate change.

3.9 Future Directions

Sustainable energy planning is a dynamic and rapidly evolving field with a wide array of potential future directions and advancements. Here are several key areas that future work in this field could build upon:

- Sustainable energy planning requires balancing trade-offs between different criteria. Current works often focus on minimizing costs or emissions, but few papers have fully integrated social acceptance or resilience to disruptions. Therefore, creating optimization algorithms that can incorporate diverse planning objectives, such as social acceptance, local community impact, and the cost of integration with existing infrastructure, is of important interest.
- Renewable energy resources like solar parks and wind farms are variable in nature. This can negatively impact grid stability, which requires new optimization techniques to handle these uncertainties. Therefore, focusing on improving optimization techniques to ensure power systems remain efficient despite fluctuations is another research interest in sustainable energy planning.
- As power grids decentralize, optimizing microgrids with renewable energy resources becomes very important. Thus, investigating algorithms for optimizing local energy consumption (e.g., peer-to-peer energy trading, demand-side management) is another future area of work, enabling efficient collaboration between various microgrids.
- Efficient storage is crucial for dealing with fluctuations in renewable energy production. Optimal sitting, sizing, and dispatch of energy storage systems are necessary to minimize costs while maintaining reliability. Therefore, exploration of optimization frameworks to model current or hybrid energy storage systems, such as combining batteries with thermal and hydrogen storage, is desired for further research.
- Policy planning requires evaluating the long-term effects of energy investments and regulations. This can involve a wide range of decision factors like carbon pricing, subsidies for renewable energy, and energy efficiency standards. In this regard, developing optimization-based models that simulate various policy scenarios to capture the outcomes of different policy interventions over time is of interest.

- The integration of electric vehicles into renewable-dominated grids adds complexity due to their variable charging demands and potential for discharging to the grid (i.e., vehicle-to-grid option). Therefore, investigating optimization algorithms that coordinate electric vehicle charging and discharging times with renewable energy production patterns is actionable. Machine learning techniques could predict energy needs, allowing for dynamic load shift optimization.
- Demand response in smart grids allows for adjusting power usage in response to supply conditions. However, optimal scheduling and strategies for shifting loads without sacrificing consumer convenience are not yet fully developed. Therefore, utilizing optimization algorithms for demand response programs that anticipate power shortages or surpluses is useful. By creating dynamic pricing models and simulating consumer behaviors, the developed optimization framework can optimize how and when households or industries adjust their energy consumption in line with renewable energy generation.

These topics represent the potential future works in sustainable energy planning. By addressing these problems, researchers and policymakers can significantly improve the efficiency, reliability, and equity of renewable energy systems while ensuring a successful transition to a low-carbon future.

3.10 Conclusion

Sustainable energy systems are crucial for mitigating environmental concerns and moving toward greener grids. Replacing fossil fuels with renewable energy is achievable, provided there are significant lifestyle changes in countries, strong international cooperation, and the implementation of various strategies. To this end, providing future proposals for sustainable planning is of paramount importance. Sustainable energy planning is inherently an interdisciplinary problem, requiring multifaceted considerations and constraints. Effective sustainable energy planning requires a balance of environmental, technical, economic, and societal considerations, making the role of optimization techniques indispensable. In this chapter, we discussed the roles of optimization techniques in achieving sustainable energy systems. The chapter presented various optimization techniques in sustainable energy planning. The optimization techniques outlined in this chapter are vital tools for meeting the challenges of modern power grid as we transition toward a more sustainable, reliable, and efficient energy future. It has been shown that how optimization methods can be seamlessly taken for a smooth transition to sustainable energy systems. Further discussions covered the advantages and disadvantages of different optimization methods. The chapter concluded that choosing an optimization technique requires knowledge about uncertainties, data availability, problem complexity, planner preferences, and so on. Finally, several directions for increasing sustainability in current power systems were presented. Shifting toward sustainability is the future choice of power and energy systems.

References

Abdullah, L., & Najib, L. (2016). Sustainable energy planning decision using the intuitionistic fuzzy analytic hierarchy process: Choosing energy technology in Malaysia. *International Journal of Sustainable Energy, 35*(4), 360–377.

Abdullah, M. A., Al-Hadhrami, T., Tan, C. W., & Yatim, A. H. (2018). Towards green energy for smart cities: Particle swarm optimization based MPPT approach. *IEEE Access, 6*, 58427–58438.

Afsordegan, A., Sánchez, M., Agell, N., Zahedi, S., & Cremades, L. (2016). Decision making under uncertainty using a qualitative TOPSIS method for selecting sustainable energy alternatives. *International Journal of Environmental Science and Technology, 13*, 1419–1432.

Al-Hamouz, Z. M., & Al-Faraj, A. S. (2003). Transmission-expansion planning based on ANON-linear programming algorithm. *Applied Energy, 76*(1–3), 169–177.

Alhumaid, Y., Khan, K., Alismail, F., & Khalid, M. (2021). Multi-input nonlinear programming based deterministic optimization framework for evaluating microgrids with optimal renewable-storage energy mix. *Sustainability, 13*(11), 5878.

Alizadeh, R., Soltanisehat, L., Lund, P. D., & Zamanisabzi, H. (2020). Improving renewable energy policy planning and decision-making through a hybrid MCDM method. *Energy Policy, 137*, 111174.

Al-Qudaimi, A., & Kumar, A. (2018). Sustainable energy planning decision using the intuitionistic fuzzy analytic hierarchy process: Choosing energy technology in Malaysia: Necessary modifications. *International Journal of Sustainable Energy, 37*(5), 436–437.

Alshammari, N., & Asumadu, J. (2020). Optimum unit sizing of hybrid renewable energy system utilizing harmony search, Jaya and particle swarm optimization algorithms. *Sustainable Cities and Society, 60*, 102255.

Arnold, U., & Yildiz, Ö. (2015). Economic risk analysis of decentralized renewable energy infrastructures—A Monte Carlo simulation approach. *Renewable Energy, 77*, 227–239.

Aryanpur, V., Atabaki, M. S., Marzband, M., Siano, P., & Ghayoumi, K. (2019). An overview of energy planning in Iran and transition pathways towards sustainable electricity supply sector. *Renewable and Sustainable Energy Reviews, 112*, 58–74.

Aviso, K. B., Lee, J.-Y., Dulatre, J. C., Madria, V. R., Okusa, J., & Tan, R. R. (2017). A p-graph model for multi-period optimization of sustainable energy systems. *Journal of Cleaner Production, 161*, 1338–1351.

Babatunde, O. M., Munda, J. L., & Hamam, Y. (2020). A comprehensive state-of-the-art survey on hybrid renewable energy system operations and planning. *IEEE Access, 8*, 75313–75346.

Baležentis, T., & Streimikiene, D. (2017). Multi-criteria ranking of energy generation scenarios with Monte Carlo simulation. *Applied Energy, 185*, 862–871.

Baziar, A., & Kavousi-Fard, A. (2013). Considering uncertainty in the optimal energy management of renewable micro-grids including storage devices. *Renewable Energy, 59*, 158–166.

Bazmi, A. A., & Zahedi, G. (2011). Sustainable energy systems: Role of optimization modeling techniques in power generation and supply—A review. *Renewable and Sustainable Energy Reviews, 15*(8), 3480–3500.

Beccali, M., Cellura, M., & Mistretta, M. (2003). Decision-making in energy planning. Application of the Electre method at regional level for the diffusion of renewable energy technology. *Renewable Energy, 28*(13), 2063–2087.

Bishop, J. D., Amaratunga, G. A., & Rodriguez, C. (2008). Using strong sustainability to optimize electricity generation fuel mixes. *Energy Policy, 36*(3), 971–980.

Bishop, J. D., Amaratunga, G. A., & Rodriguez, C. (2010). Linking energy policy, electricity generation and transmission using strong sustainability and co-optimization. *Electric Power Systems Research, 80*(6), 633–641.

Bokhoree, C., & Balnac, K. (2009). Forecasting demand for electricity in Mauritius using support vector machines for sustainable energy planning. *Journal of Environmental, 5*, 271.

Budak, G., Chen, X., Celik, S., & Ozturk, B. (2019). A systematic approach for assessment of renewable energy using analytic hierarchy process. *Energy, Sustainability and Society, 9*, 1–14.

Chang, C.-T. (2015). Multi-choice goal programming model for the optimal location of renewable energy facilities. *Renewable and Sustainable Energy Reviews, 41*, 379–389.

Chien, F., Kamran, H. W., Albashar, G., & Iqbal, W. (2021). Dynamic planning, conversion, and management strategy of different renewable energy sources: A sustainable solution for severe energy crises in emerging economies. *International Journal of Hydrogen Energy, 46*(11), 7745–7758.

Chu, S., Cui, Y., & Liu, N. (2017). The path towards sustainable energy. *Nature Materials, 16*(1), 16–22.

Clímaco, J. (2012). *Multicriteria Analysis: Proceedings of the XIth International Conference on MCDM, August 1–6, 1994, Coimbra, Portugal.* Springer Science & Business Media.

Constante-Flores, G. E., & Illindala, M. S. (2018). Data-driven probabilistic power flow analysis for a distribution system with renewable energy sources using Monte Carlo simulation. *IEEE Transactions on Industry Applications, 55*(1), 174–181.

Cosic, A., Stadler, M., Mansoor, M., & Zellinger, M. (2021). Mixed-integer linear programming based optimization strategies for renewable energy communities. *Energy, 237*, 121559.

Debnath, K. B., & Mourshed, M. (2018). Forecasting methods in energy planning models. *Renewable and Sustainable Energy Reviews, 88*, 297–325.

Demirtas, O. (2013). Evaluating the best renewable energy technology for sustainable energy planning. *International Journal of Energy Economics and Policy, 3*(4), 23–33.

DOE. (2006). *Benefits of demand response in electricity markets and recommendations for achieving them* [US Department of Energy, Technical Report]. [Online]. Available: https://www.energy.gov/sites/prod/files/oeprod/DocumentsandMedia/DOEBenefitsofDemandResponseinElectricityMarketsandRecommendationsforAchievingThemReporttoCongress.pdf

Dong, C., Huang, G., Cai, Y., Cheng, G., & Tan, Q. (2016). Bayesian interval robust optimization for sustainable energy system planning in Qiqihar City, China. *Energy Economics, 60*, 357–376.

Energy Institute—Statistical Review of World Energy. (2024). *Global primary energy consumption by source* [Technical Report]. [Online]. Available: https://ourworldindata.org/search?q=global%20primary%20energy%20consumption%20by%20source

Ervural, B. C., Evren, R., & Delen, D. (2018). A multi-objective decision-making approach for sustainable energy investment planning. *Renewable Energy, 126*, 387–402.

Ezbakhe, F., & Pérez-Foguet, A. (2021). Decision analysis for sustainable development: The case of renewable energy planning under uncertainty. *European Journal of Operational Research, 291*(2), 601–613.

Fara, L., Diaconu, A., Craciunescu, D., & Fara, S. (2021). Forecasting of energy production for photovoltaic systems based on ARIMA and ANN advanced models. *International Journal of Photoenergy, 2021*(1), 6777488.

Fathipour, F., & Saidi-Mehrabad, M. (2018). A multi-objective energy planning considering sustainable development by a topsis based augmented e-constraint. *Journal of Renewable and Sustainable Energy, 10*(3).

Giatrakos, G., Tsoutsos, T., Mouchtaropoulos, P., Naxakis, G., & Stavrakakis, G. (2009). Sustainable energy planning based on a stand-alone hybrid renewable energy/hydrogen power system: Application in Karpathos Island, Greece. *Renewable Energy, 34*(12), 2562–2570.

Gitizadeh, M., Kaji, M., & Aghaei, J. (2013). Risk based multiobjective generation expansion planning considering renewable energy sources. *Energy, 50*, 74–82.

Gómez, J., Chicaiza, W. D., Escaño, J. M., & Bordons, C. (2023). A renewable energy optimisation approach with production planning for a real industrial process: An application of genetic algorithms. *Renewable Energy, 215*, 118933.

Groumpos, P., Khouzam, K., & Khouzam, L. (1996). A dynamic programming approach to the energy management problem of photovoltaic power systems. In *Conference Record of the Twentieth IEEE Photovoltaic Specialists Conference* (pp. 1164–1167). IEEE.

Hahn, W. (2015). Making decisions with multiple criteria: A case in energy sustainability planning. *EURO Journal on Decision Processes, 3*(1–2), 161–185.

Hemmati, R., Hooshmand, R.-A., & Khodabakhshian, A. (2013). Comprehensive review of generation and transmission expansion planning. *IET Generation, Transmission & Distribution, 7*(9), 955–964.

Hemmati, S., Ghaderi, S., & Ghazizadeh, M. (2018). Sustainable energy hub design under uncertainty using benders decomposition method. *Energy, 143*, 1029–1047.

Hosseini, S. M., Carli, R., & Dotoli, M. (2020). Robust optimal energy management of a residential microgrid under uncertainties on demand and renewable power generation. *IEEE Transactions on Automation Science and Engineering, 18*(2), 618–637.

Huang, Y., Jin, M., Xie, J., Peng, Y., & Zhong, J. (2024). Dynamic Bayesian game optimization method for multi-energy hub systems with incomplete load information. *Energy, 301*, 131622.

Huang, Y., Wang, W., & Hou, B. (2019). A hybrid algorithm for mixed integer nonlinear programming in residential energy management. *Journal of Cleaner Production, 226*, 940–948.

Huo, Y., Bouffard, F., & Joós, G. (2021). Decision tree-based optimization for flexibility management for sustainable energy microgrids. *Applied Energy, 290*, 116772.

Hwang, C.-L., & Yoon, K. (1981). *Multiple attribute decision making: Methods and applications*. Springer.

Jayasekara, N., Wolfs, P., & Masoum, M. A. (2014). An optimal management strategy for distributed storages in distribution networks with high penetrations of PV. *Electric Power Systems Research, 116*, 147–157.

Jun, Z., Junfeng, L., Jie, W., & Ngan, H. (2011). A multi-agent solution to energy management in hybrid renewable energy generation system. *Renewable Energy, 36*(5), 1352–1363.

Kalogirou, S. A. (2000). Applications of artificial neural-networks for energy systems. *Applied Energy, 67*(1–2), 17–35.

Katsigiannis, Y. A., Georgilakis, P. S., & Karapidakis, E. S. (2012). Hybrid simulated annealing–Tabu search method for optimal sizing of autonomous power systems with renewables. *IEEE Transactions on Sustainable Energy, 3*(3), 330–338.

Khakpoor, M., Jafari-Nokandi, M., & Abdoos, A. A. (2017a). Dynamic generation and transmission expansion planning in the power market–based on a multiobjective framework. *International Transactions on Electrical Energy Systems, 27*(9), e2353.

Khakpoor, M., Jafari-Nokandi, M., & Akbar Abdoos, A. (2017). A new hybrid GA-fuzzy optimization algorithm for security-constrained based generation and transmission expansion planning in the deregulated environment. *Journal of Intelligent & Fuzzy Systems, 33*(6), 3789–3803.

Khatib, T., Mohamed, A., & Sopian, K. (2012). Optimization of a PV/wind micro-grid for rural housing electrification using a hybrid iterative/genetic algorithm: Case study of Kualaterengganu, Malaysia. *Energy and Buildings, 47*, 321–331.

Klemm, C., & Wiese, F. (2022). Indicators for the optimization of sustainable urban energy systems based on energy system modeling. *Energy, Sustainability and Society, 12*(1), 3.

Lazzari, F., Mor, G., Cipriano, J., Solsona, F., Chemisana, D., & Guericke, D. (2023). Optimizing planning and operation of renewable energy communities with genetic algorithms. *Applied Energy, 338*, 120906.

Lei, Y., Wang, D., Jia, H., Li, J., Chen, J., Li, J., & Yang, Z. (2021). Multi-stage stochastic planning of regional integrated energy system based on scenario tree path optimization under long-term multiple uncertainties. *Applied Energy, 300*, 117224.

Lim, J. Y., How, B. S., Rhee, G., Hwangbo, S., & Yoo, C. K. (2020). Transitioning of localized renewable energy system towards sustainable hydrogen development planning: P-graph approach. *Applied Energy, 263*, 114635.

Liu, Z., Wang, S., Lim, M. Q., Kraft, M., & Wang, X. (2021). Game theory-based renewable multi-energy system design and subsidy strategy optimization. *Advances in Applied Energy, 2*, 100024.

Lujano-Rojas, J. M., Dufo-López, R., & Bernal-Agustín, J. L. (2013). Probabilistic modelling and analysis of stand-alone hybrid power systems. *Energy, 63*, 19–27.

Ma, J., Silva, V., Belhomme, R., Kirschen, D. S., & Ochoa, L. F. (2013). Evaluating and planning flexibility in sustainable power systems. In *2013 IEEE Power & Energy Society General Meeting* (pp. 1–11). IEEE.

Manjong, N. B., Oyewo, A. S., & Breyer, C. (2021). Setting the pace for a sustainable energy transition in Central Africa: The case of Cameroon. *IEEE Access, 9*, 145435–145458.

Marzband, M., Ghadimi, M., Sumper, A., & Domínguez-García, J. L. (2014). Experimental validation of a real-time energy management system using multi-period gravitational search algorithm for microgrids in islanded mode. *Applied Energy, 128*, 164–174.

Mavromatidis, G., Orehounig, K., & Carmeliet, J. (2018). Design of distributed energy systems under uncertainty: A two-stage stochastic programming approach. *Applied Energy, 222*, 932–950.

Montajabiha, M. (2016). An extended promethe ii multi-criteria group decision making technique based on intuitionistic fuzzy logic for sustainable energy planning. *Group Decision and Negotiation, 25*, 221–244.

Nakabi, T. A., & Toivanen, P. (2021). Deep reinforcement learning for energy management in a microgrid with flexible demand. *Sustainable Energy, Grids and Networks, 25*, 100413.

Østergaard, P. A., & Sperling, K. (2014). Towards sustainable energy planning and management. *International Journal of Sustainable Energy Planning and Management, 1*, 1–6.

Otić, G., Momčilović, O., Radovanović, L., Jovanov, G., Radosav, D., & Pekez, J. (2021). Mathematical analysis of criteria for maintenance of technical systems in the function of achieving sustainability. *Sustainability, 13*(4), 1680.

Pages, A., Gondzio, J., & Nabona, N. (2009). Warmstarting for interior point methods applied to the long-term power planning problem. *European Journal of Operational Research, 197*(1), 112–125.

Paliwal, P., Webber, J. L., Mehbodniya, A., Haq, M. A., Kumar, A., & Chaurasiya, P. K. (2022). Multi-agent-based approach for generation expansion planning in isolated micro-grid with renewable energy sources and battery storage. *The Journal of Supercomputing, 78*(17), 18497–18523.

Pask, F., Lake, P., Yang, A., Tokos, H., & Sadhukhan, J. (2017). Sustainability indicators for industrial ovens and assessment using fuzzy set theory and Monte Carlo simulation. *Journal of Cleaner Production, 140*, 1217–1225.

Perera, A., & Kamalaruban, P. (2021). Applications of reinforcement learning in energy systems. *Renewable and Sustainable Energy Reviews, 137*, 110618.

Potrč, S., Čuček, L., Martin, M., & Kravanja, Z. (2021). Sustainable renewable energy supply networks optimization—The gradual transition to a renewable energy system within the European union by 2050. *Renewable and Sustainable Energy Reviews, 146*, 111186.

Qin, C., Yan, Q., & He, G. (2019). Integrated energy systems planning with electricity, heat and gas using particle swarm optimization. *Energy, 188*, 116044.

Rastgou, A., Moshtagh, J., & Bahramara, S. (2018). Probabilistic power distribution planning using multi-objective harmony search algorithm. *Journal of Operation and Automation in Power Engineering, 6*(1), 111–125.

Ratanakuakangwan, S., & Morita, H. (2021). Hybrid stochastic robust optimization and robust optimization for energy planning—A social impact-constrained case study. *Applied Energy, 298*, 117258.

Rizzo, G., Savino, G., et al. (2012). A linear programming model for the optimal assessment of sustainable energy action plans. *Proceedings of ECOS, 2012*, 398–401.

Saaty, T. L. (1980). *The analytic hierarchy process: Planning, priority setting, resource allocation*. McGraw-Hill International Book Company.

Seifi, H., & Sepasian, M. S. (2011). *Electric power system planning: Issues, algorithms and solutions* (Vol. 49). Springer.

Solangi, Y. A., Tan, Q., Mirjat, N. H., & Ali, S. (2019). Evaluating the strategies for sustainable energy planning in Pakistan: An integrated SWOT-AHP and fuzzy-TOPSIS approach. *Journal of Cleaner Production, 236*, 117655.

Stoeglehner, G., Niemetz, N., & Kettl, K.-H. (2011). Spatial dimensions of sustainable energy systems: New visions for integrated spatial and energy planning. *Energy, Sustainability and Society, 1*, 1–9.

Swain, R. B., & Karimu, A. (2020). Renewable electricity and sustainable development goals in the EU. *World Development, 125*, 104693.

Szabó, S., Bódis, K., Huld, T., & Moner-Girona, M. (2013). Sustainable energy planning: Leapfrogging the energy poverty gap in Africa. *Renewable and Sustainable Energy Reviews, 28*, 500–509.

Tsangas, M., Papamichael, I., & Zorpas, A. A. (2023). Sustainable energy planning in a new situation. *Energies, 16*(4), 1626.

Tsangas, M., & Zorpas, A. A., Jeguirim, M. (2022). Sustainable renewable energy policies and regulations, recent advances, and challenges, In *Renewable energy production and distribution* (pp. 449–465).

Turgut, Z. K., & Tolga, A. C. (2018). Fuzzy MCDM methods in sustainable and renewable energy alternative selection: Fuzzy vikor and fuzzy todim, In *Energy management—Collective and computational intelligence with theory and applications* (pp. 277–314).

Vera, E. G., Canizares, C. A., Pirnia, M., Guedes, T. P., & Trujillo, J. D. M. (2022). Two-stage stochastic optimization model for multi-microgrid planning. *IEEE Transactions on Smart Grid, 14*(3), 1723–1735.

Wang, G., Chao, Y., Cao, Y., Jiang, T., Han, W., & Chen, Z. (2022). A comprehensive review of research works based on evolutionary game theory for sustainable energy development. *Energy Reports, 8*, 114–136.

Wen, M., Zhou, C., & Konstantin, M. (2023). Deep neural network for predicting changing market demands in the energy sector for a sustainable economy. *Energies, 16*(5), 2407.

Xydis, G., & Koroneos, C. (2012). A linear programming approach for the optimal planning of a future energy system. Potential contribution of energy recovery from municipal solid wastes. *Renewable and Sustainable Energy Reviews, 16*(1), 369–378.

Yanık, S., Sürer, Ö., & Öztayşi, B. (2016). Designing sustainable energy regions using genetic algorithms and location-allocation approach. *Energy, 97*, 161–172.

Yu, J., Ryu, J.-H., & Lee, I.-B. (2019). A stochastic optimization approach to the design and operation planning of a hybrid renewable energy system. *Applied Energy, 247*, 212–220.

Zadeh, L. A. (1965). Fuzzy sets. *Information and Control, 8*(3), 338–353.

Zadehbagheri, M., Kiani, M. J., Pirouzi, S., Movahedpour, M., & Mohammadi, S. (2023). The impact of sustainable energy technologies and demand response programs on the hub's planning by the practical consideration of tidal turbines as a novel option. *Energy Reports, 9*, 5473–5490.

Zhang, W., Maleki, A., Rosen, M. A., & Liu, J. (2018). Optimization with a simulated annealing algorithm of a hybrid system for renewable energy including battery and hydrogen storage. *Energy, 163*, 191–207.

Chapter 4
Cutting-Edge Technologies in Sustainable Energy Planning and Decarbonization

Mohammad Mehdi Amiri, Mohammad Taghi Ameli, Hossein Ameli, Mohammad Reza Aghamohammadi, and Goran Strbac

Abstract It is impossible to overstate the importance of cutting-edge technologies in a multitude of domains. These technologies are applied across a range of sectors and applications, including those related to emission reduction and energy efficiency. The implementation of cutting-edge technologies, including advanced grid management systems, energy storage solutions, and smart grids (SGs), has facilitated the seamless integration of renewable energy sources into the existing energy infrastructure. These developments are crucial for transitioning toward a more sustainable and environmentally friendly energy future, offering significant advantages for consumers and the natural environment. The deployment of these technologies will facilitate a reduction in the carbon footprint, encourage the sustainable utilization of resources, and contribute to biodiversity protection, among other environmental benefits. This chapter will investigate the concept of sustainable energy, outlining its objectives and examining the aforementioned cutting-edge technologies in detail. Subsequent to their categorization, comprehensive explanations of their mechanisms, applications, and prospective impact on sustainable energy development will be provided. Furthermore, this chapter will examine the emerging challenges and trends associated with these technologies, including scalability, cost-effectiveness, regulatory barriers, and technological advancements. Additionally, it will discuss the new trends in these technologies and their future outlook. This book chapter highlights the significance of cutting-edge technologies and their role in the development of sustainable energy

M. M. Amiri (✉) · M. T. Ameli · M. R. Aghamohammadi
Faculty of Electrical Engineering, Shahid Beheshti University, Tehran, Iran
e-mail: moh_amiri@sbu.ac.ir

M. T. Ameli
e-mail: m_ameli@sbu.ac.ir

M. R. Aghamohammadi
e-mail: m_aghamohammadi@sbu.ac.ir

H. Ameli · G. Strbac
Control and Power Group, Imperial College London, London, UK
e-mail: h.ameli14@imperial.ac.uk

G. Strbac
e-mail: g.strbac@imperial.ac.uk

I. Rahimi et al. (eds.), *Sustainable Supply Chain of Renewable Energy Networks*, Energy Systems in Electrical Engineering, https://doi.org/10.1007/978-981-96-4940-2_4

Keywords Cutting-edge technologies · Sustainable energy · Carbon emission · Direct air capture · Carbon capture · Utilization and storage · Blockchain · Smart grids · Hydrogen storages

4.1 The Main Objectives of Sustainable Energy

The main objectives of sustainable energy planning are centered on achieving a balance between meeting current energy needs and ensuring that future generations can meet theirs. These objectives focus on environmental, economic, and social dimensions of sustainability. Here are the primary objectives:

- Environmental protection
- Economic viability
- Social equity
- Technological innovation
- Policy and governance
- Efficiency and conservation.

Sustainable energy planning seeks to strike a balance between current and future energy needs with respect to environmental, economic, and social sustainability. The main goals involve the reduction of greenhouse gas emissions, saving resources, and less pollution for the protection of the environment (Gunnarsdóttir et al., 2021). Economically, it looks at ways to establish cost-effective and secure energy systems, generate jobs, and guarantee energy security (Herc et al., 2022). This includes social equity: universal access to affordable energy, communities involved in the planning process, and improved public health due to a reduction in pollution-related illnesses (Sadiqa, 2023).

Technological innovation, policy, and efficiency are considered some of the key pillars for sustainable energy planning. Investment in research and development drives the development of scalable and adaptable technologies, thereby laying down the foundation for a more efficient adoption of renewable energy into existing systems. Policy frameworks, financial incentives, and international co-operation support such efforts. Further, energy efficiency improvements along with demand management through advanced technologies of SGs and demand response systems avoid wastage of energy and enhance the sustainability of the overall system (Nastasi et al., 2022).

By focusing on these objectives, sustainable energy aims to create a resilient, inclusive, and environmentally friendly energy system that supports long-term ecological balance, economic stability, and social well-being. This chapter will investigate a range of key technologies promising for enhancing sustainability and environmental management. It includes hydrogen technologies, which highlight the production of green hydrogen and hydrogen fuel cells as a renewable energy source: blockchains for energy transactions, carbon capture, utilization and storages (CCUS), which covers direct air capture (DAC) and carbon utilization efforts, etc. Each technology

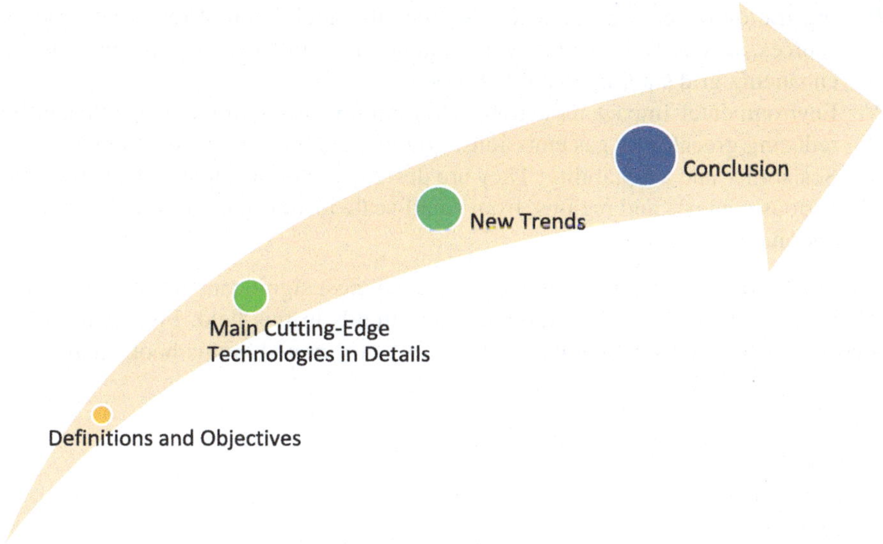

Fig. 4.1 General pathway of the book chapter

is explored in detail, addressing potential benefits, challenges, and emerging trends, underscoring their significant role in transitioning toward a cleaner, more sustainable energy future. Figure 4.1 shows the general structure of this chapter.

4.2 Cutting-Edge Technologies

Cutting-edge technologies refer to the most advanced, innovative, and pioneering developments in a particular field at any given time. These technologies represent the forefront of progress and often incorporate the latest scientific and engineering advancements. In an academic context, cutting-edge technologies are defined as those that embody the latest advancements in research and development within a specific scientific or engineering discipline. These technologies are typically at the leading edge of innovation and have undergone rigorous testing, validation, and peer review to ensure their efficacy and reliability. Cutting-edge technologies in sustainable energy planning aim to reduce greenhouse gas emissions, increase energy efficiency, and promote renewable energy sources. Key characteristics of cutting-edge technologies in this field include (Shah et al., 2022; Sharipov et al., 2022):

I. Innovation and advancement: These technologies represent the forefront of research and development in sustainable energy, incorporating recent scientific discoveries and engineering breakthroughs.
II. Efficiency improvement: They significantly improve the efficiency of energy systems, reducing losses, and optimizing the use of resources.

III. Integration of renewables: They facilitate the integration of renewable energy sources, such as solar, wind, hydro, and biomass, into existing energy systems, enhancing grid stability and reliability.
IV. Environmental impact reduction: They minimize environmental impacts by reducing greenhouse gas emissions, pollution, and resource depletion.
V. Scalability and adaptability: They are designed to be scalable and adaptable to various contexts and regions, from small-scale local installations to large-scale national grids.

In the following section, we will examine the most significant cutting-edge technologies, and finally, we will introduce new trends in this field. Five of the most important cutting-edge technologies that will be examined in this book chapter are as follows:

A. Green Energy Storage Solution
B. Smart Grids
C. Biofuels
D. Blockchain
E. CCUS.

4.3 Cutting-Edge Green Energy Storage Solution

In an effort to create a more sustainable and eco-friendly future, green energy storage solutions, such as solar, wind, and hydroelectric power, have become increasingly popular. This type of clean energy is characterized by intermittency. A sustainable energy landscape requires green energy storage solutions to address this challenge. Green energy storage is often referred to as renewable energy storage or sustainable energy storage. It involves capturing and storing energy generated from renewable and environmentally friendly sources. Solar and wind power, for instance, produce energy intermittently, meaning they generate electricity only when the sun shines or when the wind blows. The use of green energy storage systems guarantees a continuous and reliable supply of energy when excess power is available (Saritha et al., 2024). While green energy storage is crucial in balancing the supply and demand of renewable energies, they also come with their share of challenges. For instance, battery storage systems require substantial amounts of rare earth metals like lithium, cobalt, and nickel, especially those using lithium-ion technology. Their mining and extraction could bring environmental degradation, habitat destruction, and pollution. Besides, the waterless energy storage massive system is quite costly, and handling the safe disposal or recycling of used batteries is a challenge, which in turn may result in a lot of future problems with waste management.

4.3.1 *Pumped Hydrostorage*

When energy demands are low, this technology pumps water uphill. An electric generator generates electricity through the release of water from a reservoir when demand is high. Hydropower remains the largest renewable source of electricity generation in the world, including pumped storage. The use of pumped hydrostorage supports the integration of renewable energy and the balancing of supply and demand (Alonso, 2024). The technical specifications of this type of storage system are as follows:

- Efficiency: Pumped hydro has high round-trip efficiency, typically around 70–80%. This means that for every unit of energy used to pump water uphill, you get back 70–80% of that energy when it's released.
- Capacity: Pumped hydrofacilities can store large amounts of energy. Some projects have capacities in the gigawatt range, making them suitable for grid-scale energy storage.
- Location: These systems require specific geographical features—a suitable elevation difference between reservoirs and available land. Countries like China, the United States, and Japan have significant pumped hydrocapacity.
- Challenges: While pumped hydro is reliable and proven, it faces challenges related to environmental impact (habitat disruption, water use) and site availability.
- Applications: Large-scale storage with long discharge durations (Elalfy et al., 2024).

4.3.2 *Compressed Air Energy Storage (CAES)*

In CAES systems, excess energy is stored underground in caverns by compressing air. Through turbines, compressed air is used to generate electricity when electricity is needed. CAES is efficient and has a long lifespan. CAES represents a promising avenue for sustainable energy storage, especially as we seek long-duration solutions to decarbonize our electricity systems (Yu et al., 2019). The technical specifications of this type of storage system are as follows:

- Large-scale storage: CAES operates on a large scale and can expand the utilization of sustainable energy sources. It can store surplus power generated from wind turbines and solar panels.
- Process: During periods of excess electricity, air is compressed and stored in underground caverns or above-ground tanks. When needed, the compressed air is released, expanding through turbines to generate electricity.
- Efficiency: CAES has the potential for high efficiency, especially when combined with advanced technologies like adiabatic or isothermal processes (Rabi et al., 2023).

- Challenges: Presently, CAES suffers from energy losses and often relies on natural gas for operation. However, ongoing research aims to improve its efficiency and environmental impact (Rabi et al., 2023).
- Applications: Large-scale storage with high capacity and long discharge times (Rabi et al., 2023).

4.3.3 Flywheel Energy Storage

When electricity is needed, kinetic energy is converted back into electricity by a spinning rotor in a flywheel energy storage system. Flywheels are known for their rapid response time and high power density. The technical specifications of this type of storage system are as follows (Li & Palazzolo, 2022):

I. Advantages: FES offers several unique advantages:
   - Long life cycles: Flywheels have long operational lifetimes.
   - High power density: They can deliver power rapidly when needed.
   - Minimal environmental impact: Flywheels are environmentally friendly.
   - Fast response and voltage stability: They provide stable power quality.

II. Operation: During periods of excess electricity, energy is fed into the rotational mass of the flywheel, storing it as kinetic energy. When demand arises, the stored energy is released.

III. Applications
   - Grid stabilization: FES can enhance grid stability by providing rapid response during fluctuations.
   - Transportation: Flywheels can be used in hybrid vehicles and trains.

IV. Recent developments: Notably, there has been progress in utility-scale deployment for the electrical grid and renewable energy applications.

4.3.4 Hydrogen Storage

The most abundant element in the universe is hydrogen, a colorless, odorless gas. For generating electricity, it can be used in fuel cells (FC). Unlike fossil fuels, hydrogen is a clean fuel that has no carbon (Amiri et al., 2024b; Dincer & Aydin, 2023). Hydrogen can be produced several ways, including through fossil fuel extraction, electrochemical methods, thermochemical cycles, photonic methods, biological methods, etc. A promising way to treat wastes while producing hydrogen is biomass valorization through thermochemical processes. Biomass is converted to liquid biofuel, solid char and syngas through the process of pyrolysis (Cheng et al., 2020). Hydrogen produced via other methods such as photocatalysis, photofermentation, dark fermentation, and supercritical water gasification has also shown to have a lower global warming potential than hydrogen produced via fossil fuels (Aydin & Dincer, 2022; Aydin et al., 2021). By producing hydrogen from waste streams, greenhouse gas emissions can be reduced. In addition to contributing to sustainable development, waste treatment contributes to a clean environment, water and sanitation, and life below water.

Hydrogen storage is a critical component in advancing sustainable energy development. It enables the efficient utilization of hydrogen as a clean fuel source. However, its low ambient temperature density poses challenges for practical applications. Effective hydrogen storage methods are essential to address these limitations. A characteristic of hydrogen that cannot be ignored is that it occupies a much larger volume than any other gas at normal atmospheric pressure, even though it is an ultralight gas (around 11 times lighter than air we breathe). A volume of 11 m^3 is required to store 1 kg of hydrogen (Hydrogen storage: a challenge for the development of the industry, https://www.terega.fr/en/our-activities/hydrogen/hydrogen-storage-a-challenge-for-the-development-of-the-industry/). Hydrogen storage methods can be divided into the following three categories (Muhammed et al., 2023):

1. Gas storage: Hydrogen can be stored physically as a gas in high-pressure tanks (350–700 bar). These tanks are suitable for onboard automotive applications.
2. Liquid storage: Cryogenic temperatures are required to store hydrogen as a liquid due to its low boiling point (− 252.8 °C).
3. Solid storage:
    - Adsorption: Hydrogen can be stored on the surfaces of solid materials.
    - Absorption: Hydrogen can be stored within solid materials.

The U.S. Department of Energy (DOE) aims to develop hydrogen storage systems that meet performance targets (Hassan et al., 2023):

- Energy density: 1.5 kWh/kg (4.5 wt% hydrogen)
- Volume density: 1.0 kWh/L (0.030 kg hydrogen/L)
- Cost: $10/kWh ($333/kg stored hydrogen capacity)

Among the challenges in this regard, high-density hydrogen storage remains a challenge for stationary, portable, and transportation applications. Researchers actively explore novel materials and engineering solutions to enhance storage capabilities (Irham et al., 2024). Recently, several exciting developments have been made in the materials of hydrogen storage. These developments have essentially been the backbone of the viable hydrogen economy in pursuit because their focus has been on making hydrogen storage more efficient and cost-effective. Following are a few key developments (Jangir et al., 2024; Rampai et al., 2024; Ren et al., 2023):

- Metal hydrides: Most of the metal hydrides like MgH_2 are being nanostructured to improve their hydrogen storage capacity and kinetics. Nanostructuring is helpful in achieving near room temperature hydrogen desorption with high capacity.
- Carbon-based materials: Extensive research is focused on CNTs and graphene as well, due to their high surface areas and good hydrogen storage capacity.

- Metal–organic frameworks (MOFs): MOFs are highly porous materials with a capacity for large volume storage of hydrogen at relatively low pressures. They are being tuned in order to further enhance their storage capacity and stability.
- Metastable alloys: New systems with metastable structures, such as nanocrystalline and amorphous alloys, are under development. Such materials exhibit improved hydrogenation kinetics and superior storage capacity.
- Zeolites and clathrates: These nanoporous materials are also under consideration due to their efficiency in hydrogen storage.

4.4 Smart Grids

Electricity has conventionally been distributed in a rigid one-way fashion through transmission and distribution systems serving large power plants. Decentralization of generation, SGs, and new energy carriers have given way to greater accessibility than the traditional grid (Kabeyi & Olanrewaju, 2022). In an ideal two-way electricity grid, big, medium, and small-scale power plants can provide grid power in mixed modes. This can ensure sustainability in the energy sector for sustainable development. A SG can be defined to represent an electricity network comprising transmission lines, substation, transformer, and other devices with digital and advanced technology for monitoring, managing, and evacuating electricity from all the power stations in a way that accommodates load variation. These SGs are capable of integrating solar panels, wind farms, and other renewable sources of energy; they also provide for distributed energy resources in such a way that there is optimal production and distribution of electricity.

SGs were fitted with sensors and communication devices that didn't only provide real-time energy usage monitoring and smart equipment control but also allowed for two-way communication among the utilities and end users/devices. With further development, more attention started to be paid to energy management systems for a smart home, such as HEMS and BEMS. As a result, with SGs, more opportunities have appeared for maximizing energy gain and consumption. In this way, all activities and operations on the entire electricity grid, from planning to generation, transmission, distribution, and end use, can be optimized. SGs use digital communication technology to detect and respond to local changes in electrical demand, making electricity distribution more reliable and efficient. In order to maximize reliability, resilience, flexibility, and stability in the electricity market, SG coordinates the needs of generators, grid operators, end users, and stakeholders in the market (Khalid, 2024).

SGs are a crucial element in the development of sustainable energy systems and the achievement of energy goals. They automatically reroute power when equipment fails, thereby reducing oscillations and overloads that can lead to blackouts (Kabeyi & Olanrewaju, 2022). They support decentralized power production, while smart meters, on the other hand, provide consumption data in real time, enabling consumers to adapt their use to reduce costs. This flexibility on the demand side optimizes the whole electricity supply system, managing generation efficiently while reducing environmental impacts. SGs also allow market integration and facilitate harmonious operation among all stakeholders by enhancing performance and the quality of provided products and services. In addition, they deliver crucial real-time information for planning stability and cost-effectiveness (Moreno Escobar et al., 2021). By enhancing reliability, security, and stability of supply, SGs enable new business models due to enhanced information flows, consumer engagement, and better system control. They can also turn consumers into prosumers who can produce, consume, and even sell energy back to the grid, thereby facilitating higher economic and operational efficiency in the energy markets (Amiri et al., 2024a).

Despite the many advantages of SGs, we can mention some examples of their negative consequences. The digital infrastructure underlying SGs—sensors, communication networks, and control systems—raises several issues related to cybersecurity and data privacy. SGs are susceptible to cyber-attacks; disruption of energy supplies is one of the potentially destabilizing consequences. Widespread deployment of smart meters and accompanying infrastructure may additionally face challenges related to public acceptance, privacy, and ownership of data. Moreover, the high deployment cost of smart grid technology may further lead to disparities between those regions able to afford modernization and those that cannot.

4.4.1 Artificial Intelligence (AI) and Internet of Things (IoT) in SGs

With the help of AI, SGs evaluate the environment and take action to maximize a given objective. In order to integrate renewable energy, stabilize energy networks, and reduce financial risks associated with infrastructure instability, AI is essential. AI can address intermittent renewable energy sources, for example, due to its self-learning, adaptability, and calculation capabilities. The "duck curve" represents an imbalance of production and consumption and makes these sources of energy challenging to control. With AI in SGs, this challenge can be addressed by rebalancing the inequity between production and consumption (Meydani et al., 2024).

4.4.1.1 The Utility Sector's Perspective on AI in Smart Energy Solutions

The use of smart grid technologies makes utilities' operations more transparent and competitive (Meydani et al., 2024). SG technologies are used for a variety of applications, including:

- Fault detection and management: AI systems can quickly identify and isolate faults in the grid, minimizing the impact of outages. This improves the overall reliability and resilience of the power grid.

- Resilience: In most cases, renewable energy produced by cooperatives or prosumers is intermittent and variable. Use of sensors and automation can detect weak parts of the grid and reroute it automatically-store in surplus during peak generation and reroute when there is a shortage (Meydani et al., 2024).
- More accurate forecasts: The shift in consumption patterns creates frequent price volatility within the utility industry. A mix of advanced metering infrastructure (AMI) and AI market forecasting does better jobs compared to what was hitherto achievable through the legacy methods by combining AMI data and AI.
- Improved automated switching: The switching protocol will soon be automated once the AI tools, while predicting the grid imbalance, can distinguish between a brief power outage and a full-on outage. Thus, utility companies can reroute the energy or isolate the affected areas before actual severe damages occur.
- More flexible demand-side management (DSM): It is very difficult for utility companies to meet peak energy demand. By using AI and smart meters in homes and offices, they will be able to plan, execute, and monitor changes in energy demand to ensure that they are able to meet those demands (Arumugham et al., 2023).
- Cybersecurity: AI enhances the security of SGs by detecting and responding to cyber threats in real-time. This is important for protecting critical infrastructure from attacks.
- Grid stability and reliability: By monitoring and adjusting grid energy distribution in real-time, SGs improve grid stability and reliability as a result of electrification of a wide range of economic activities (Moreno Escobar et al., 2021).

4.4.1.2 Applications of AI in Smart Energy Solutions: The Consumer Perspective

Below are some of the ways smart grid technologies can help to improve customer satisfaction (Moreno Escobar et al., 2021):

- Lower costs: Customers can get hourly energy usage assessments powered by AI with SG management and smart metering, helping them understand where and when they use the most energy, as well as optimize their typical daily routines to lower peak usage.
- Improved sustainability and transparency: The use of smart grid data can increase customer engagement and democratize the grid, giving them a better understanding of how their energy is produced and allowing them to make more responsible choices.
- Fewer outages: Both residential and commercial customers can benefit from AI tools in reducing outages and mitigating their impact. With the rising threat of brownouts and other disruptions caused by weather events and record temperatures, consumers will feel more secure and confident.

4.4.1.3 The Trade-Offs Between Centralized and Decentralized AI Models for SGs Management

The balance of resilience, efficiency, and scalability of AI models lies in the trade-offs between the centralized and decentralized models for smart grid management. Centralized AI models have the advantage of a unified control structure for optimized decisions on the whole grid. This simplifies the integration of diverse data sources and may enable more powerful computation-hence, potentially faster and more accurate response to grid dynamics. Centralized models, however, present the risk of single failures and may subsequently increase problems with scalability as DERs expand outward. With decentralized AI models, many think in terms of operation at the edge of the grid, offering better resilience by distribution of intelligence across local nodes. In that respect, such decentralization might reduce the risk of a systemic failure and also provide better real-time responsiveness because processing is closer to the source of data creation. However, for decentralized models, this might be limited to only a small extent since the coordination among distributed agents can become very complex and less effective. Thus, centralized versus decentralized AI in SGs is a tradeoff between global optimization/ control requirements and flexibility, scalability, and adaptability enabled by decentralization, against an increasingly dynamic and decentralized energy landscape (Mostafa & Saad, 2024; Zibaeirad et al., 2024).

4.5 Biofuels

In recent years, increasing concerns about climate change, energy security, and sustainability have translated into growing interest in biofuels. Accordingly, numerous benefits of biofuels over traditional fossil fuels include the following: lower greenhouse gas emissions, reduced foreign oil dependence, and better rural economies. A biofuel is also locally producible and sourced from renewable feed-stocks, making it more sustainable because long-distance fuel transport is no longer a necessity. It is conveyed that biofuels have the possibility of being used with existing infrastructure without major changes, thus making them a good alternative to fossil fuels. Production and consumption of these crops involve very complex generation implications that are ethical, social, and economic in nature, which have to be taken into consideration if both sustainable and equitable economic growth and reduction of climate change effects are to be achieved. For example, there arises a tug in food and land use with more demands for food and fuel; this, to another account, drives insecurity in food distribution and affects the biodiversity of a place. Besides these, biofuels are costly in capital investment and yield low and uncertain returns on investments, which makes it an unattractive economic activity (Nath, 2024). Various types of Biofuels used worldwide are as follows:

- Wood
- Biogas
- Biodiesel
- Ethanol
- Methanol
- Butanol.

4.5.1 Economic and Environmental Impact of Biofuels

Biofuels are considered carbon–neutral energy sources because carbon dioxide (CO_2) is absorbed by photosynthesis during the plant growth stage. In this regard, alternatives to petroleum resources, such as biofuels, are becoming increasingly important due to their environmental friendliness and sustainability. The use of these fuels helps reduce vehicle emissions and mitigate environmental change. Biofuels present a complex mix of benefits and challenges, making it essential to consider both economic and environmental factors when evaluating their overall impact (Demırbas, 2017).

4.5.1.1 Economic Impact

The economic impact of biofuels presents both positive aspects and challenges. On the positive side, biofuel demand can significantly boost agricultural income by increasing the prices and demand for crops used as feedstocks. This, in turn, can lead to enhanced income for farmers. Furthermore, biofuels contribute to national energy security by reducing dependence on imported petroleum, offering a more stable and secure energy supply. However, several challenges accompany these benefits. Biofuels often rely on government subsidies and market interventions to compete with fossil fuels, which can result in economic inefficiencies. Additionally, the increased demand for biofuel feedstocks can drive up food prices, creating challenges for food security and affordability.

4.5.1.2 Environmental Impact

From an environmental perspective, biofuels have the potential to reduce greenhouse gas emissions compared to fossil fuels, especially when considering their lifecycle emissions. Being produced from renewable resources, biofuels offer a sustainable alternative that, unlike fossil fuels, can be replenished indefinitely. However, challenges exist here as well. A shift in land use patterns, increased water use, and even pollution can result from the production of biofuels. Biofuels can emit more greenhouse gases than fossil fuels on an energy-equivalent basis depending on their feedstock and production methods, complicating their environmental benefits.

4.5.2 Implicit Profitable Benefits of Biofuel Product

Following implicit benefits highlight the multifaceted value of biofuels, extending beyond direct financial gains to include broader economic and environmental advantages (Biofuels: a sustainable solution for a green energy future, https://www.spectrafuels.com/biofuels-a-sustainable-solution-for-a-green-energy-future/; Malik et al., 2024). The production of biofuels brings several socio-economic and technological benefits. Job creation is one such advantage, particularly in rural areas, where biofuel production can generate employment opportunities from farming to processing, thereby stimulating local economies. Additionally, investment in biofuel research and development can foster innovation and create high-tech jobs, contributing to economic growth.

Another key benefit is energy independence. By producing biofuels domestically, countries can reduce their reliance on imported oil, saving on import costs and improving trade balances. This shift toward local energy production also supports market diversification. Biofuels provide an alternative market for agricultural products, helping stabilize prices and reducing the volatility associated with traditional food markets. Environmental services also play a role in the economic benefits of biofuels. Companies engaged in biofuel production may earn carbon credits or other environmental incentives, offering financial advantages. Moreover, the utilization of agricultural or industrial waste for biofuel production can lower disposal costs while creating additional revenue streams. Lastly, biofuel production drives technological advancements. Innovation in biofuel technology can have spillover effects, leading to improvements in related sectors such as agriculture and waste management, further broadening the economic and environmental benefits.

4.5.3 Implicit Profitable Disbenefits and Impacts of Biofuel Product

Following implicit disbenefits highlight the need for careful consideration and management of biofuel production to balance economic, environmental, and social factors (Biofuels: a sustainable solution for a green energy future, https://www.spectrafuels.com/biofuels-a-sustainable-solution-for-a-green-energy-future/; Liu et al., 2021).

A. *Resource competition*

- Food versus fuel: The production of the new biofuels from crops may well compete with food production, which will drive up food prices and lead to eventual food insecurity.
- Land use: Large-scale production of biofuels can result in lands being cleared of their natural habitats for agricultural use, leading to deforestation and loss of biodiversity.

B. *Environmental degradation*

- Problem of Water Use: Biofuel crops require much water and can use up the local water supply to such an extent that it may affect agriculture in those particular areas.
- Soil health: Intensive farming for feedstock used in the production of biofuels can lead to soil degradation and, over the long term, lower agricultural productivity.

C. *Economic volatility*

- Market fluctuations: Being subject to oil price, government policy, and crop yield changes, the markets for biofuels are generally quite unpredictable-a set of factors that could affect profitability.
- Dependence on subsidy: Biofuel industries rely heavily on subsidies from the government to fend off competition, and changes in policy may make some of these uneconomically viable.

D. *Energy balance*

- Net energy gain: Some biofuels have a low net energy gain-that is, the energy it takes to make them is not much less than the energy they produce.
- Lifecycle emissions: Some biofuels may not necessarily cut greenhouse gas emissions that much compared to fossil fuels, depending on the process for production.

E. *Social impacts*

- Displacement: Biofuel large-scale projects displace the local people and drift them away from their traditional modes of livelihood.
- Labor conditions: In most developing countries where biofuels are produced, poor labor conditions affect the welfare of the workers.

4.5.4 Recent Advancements in Global Biofuel Production

This section of the book chapter will focus on advances and innovations in biofuel technology that have the potential to improve the sustainability of biofuel production. Over the past decade, there has been notable progress in energy recovery from biomass and the use of biofuels as a renewable alternative to fossil fuels.

One area of innovation is genetic engineering and synthetic biology. Through these techniques, scientists can modify the genetic makeup of microorganisms and plants to increase biomass yield, enhance resistance to environmental stresses, and optimize their capacity to convert biomass into biofuels. This holds the potential to improve the quality of raw materials, increase fuel yields, and lower production costs. Algae-based biofuel is another promising area. Algae are considered ideal feedstocks due to their high oil content and rapid growth. Advances in algal cultivation, genetic modification, and extraction techniques are helping to improve the efficiency of algae-based biofuel production. Additionally, the use of algae on non-agricultural land and in wastewater systems minimizes competition with food production and reduces environmental impacts, offering a sustainable approach to bioenergy.

The conversion of waste into energy also presents significant potential. Agricultural residues, food waste, and forest residues can be transformed into biofuels, providing a solution for both waste management and energy generation. Recent innovations in waste-to-energy technologies include biochemical and thermochemical processes such as anaerobic digestion, gasification, and pyrolysis, all of which are effective methods for converting various types of waste into biofuels, biogas, or syngas (Sundaram et al., 2023). Integrated biorefineries represent another important advancement in biofuel production. These facilities use diverse biomass feedstocks to produce a range of products, including biofuels, chemicals, and materials. By integrating different transformation processes—such as biochemical, thermochemical, and catalytic technologies—into one facility, integrated biorefineries maximize resource efficiency, diversify product outputs, and enhance the economic viability of biofuel production. This approach offers a promising path toward more sustainable and comprehensive biofuel systems (Prasad et al., 2024; Sundaram et al., 2023).

4.6 Blockchain Technology

With each passing day, it is becoming evident that blockchain technology might prove to be a game changer in every industry around the world—including the sustainable energy market. Blockchain technology ensures that transactions take place in a secure and utterly transparent manner sans intermediaries' interference. This lays the path to very transparent and secure tracking and verification of renewable energies production and consumption, which might change the renewable energy industry. Blockchain technology allows decentralized SGs through the utilization of DERs such as solar panels and windmills. Because of its platforms, the efficiency of DERs to sell excess energy can be realized, hence making energy trading reliable. Energy buying and selling contracts could be automated using smart contracts, which reduces transaction costs and times of settlement. In gas and energy commodity trading, blockchain reduces risks in exploitation, adding to the transparency by recording and tracking energy data (Taherdoost, 2024).

One of the ways it contributes to environmental sustainability is the application of blockchain to carbon credit markets, which can tokenize and digitally distribute environmental assets, showing the rising growth in these markets and leading international bodies and organizations involved therein. Due to the nature of blockchain, which is decentralized, there is a great and transparent system where sustainability is promoted. The core attribute of a distributed ledger has the potential to change many industries, ranging from finance and food to energy and environmental projects, for good. Next section will discuss blockchain technology for helping the development of sustainable energy. It is important to note that decentralization is only one aspect of blockchain technology. A blockchain network's environmental footprint is heavily influenced by the consensus mechanisms responsible for authenticating transactions and maintaining integrity. Proof of Work (PoW) and Proof of Stake (PoS) exhibit stark differences in energy usage.

- ***Proof of Work (PoW)***

  PoW is a consensus mechanism known for its high energy consumption. It requires miners to solve complex mathematical puzzles, which demand significant computational power and electricity. For example, Bitcoin, the most prominent blockchain using PoW, consumes energy levels comparable to those of entire countries. Despite its energy demands, PoW is highly secure, as altering the blockchain would require an immense amount of computational effort, making it highly resistant to attacks.
- ***Proof of Stake (PoS)***

  On the other hand, PoS offers a much more energy-efficient alternative to PoW. Instead of relying on energy-intensive puzzle-solving, PoS selects validators based on the number of tokens they hold and are willing to "stake" as collateral. This reduces energy consumption significantly. PoS also maintains a high level of security, though it operates on economic incentives rather than computational power to ensure the blockchain's integrity.

4.6.1 Renewable Energy Trading

Along with this, blockchain technology is now allowing peer-to-peer transactions in the renewable energy sector by earmarking clean energy exchange. With this innovative idea, home residents would exchange renewable energy certificates for the purchase and sale of renewable energy. A number of renewable energy projects have already shown the use of such applications. In India's Uttar Pradesh, for instance, blockchain solutions have already allowed homeowners to trade surplus solar electricity with other users in their grid network. Applications such as these from Indian homeowners have therefore been in a position to act as proof of concept regarding the functionality of blockchains in regional renewable energy markets (Rejeb et al., 2024; Taherdoost, 2024).

By making energy transactions transparent and tamper-proof, blockchain allows for real-time monitoring. Due to its decentralized nature, this system eliminates the need for intermediaries, which reduces transaction fees and enhances vendor-purchaser trust (Taherdoost, 2024).

4.6.2 Sustainable Supply Chains

Sustainability, other than environmental causes, is ethical procurement, giving fair working conditions, and creating sustainable economic systems. In such dimensions of sustainability, blockchain technology should be implemented to keep a sharp eye on products in the production cycle, supply, usage, and recycling. We can do this by providing information about the origin of products, transportation mode, and sustainability credentials (Cui et al., 2023). A set of ways can be achieved by blockchain technology capabilities.

- Maintaining fair employment standards.
- Achieving a more equitable distribution of profits.
- Enabling the supply chain ecosystem to be traceable and authenticated, thereby eradicating human trafficking and other illegal practices.
- Reliable monitoring of a product's carbon footprint.
- Carbon emissions taxes, for example, could be integrated into product pricing mechanisms.

4.6.3 Carbon Credits and Emissions Tracking

Blockchain-based smart contracts make tracking carbon production much easier, and records related to environmental activity are guaranteed to be authentic and tamper-proof because of blockchain's immutable nature. In this respect, leading-edge technology enables companies, by creating zero-knowledge proofs, to indefensibly confirm without necessarily giving out proprietary information their declaration of emission standards and levels. If carbon credits are tokenized with the use of blockchain technology, then it already supports fractional ownership and opens up toward a wider audience by creating an ecosystem around the global market of those environmentally vital assets (Effah et al., 2021).

4.6.4 Challenges

While blockchain technology brings in aspects of transparency and decentralization in energy management, it is that at very high energy consumption, especially in the case of PoW. The computational power behind blockchain mining results in great electricity use, which can negate all the goals of its contribution toward sustainability. The second point is that, by nature, blockchain is difficult to regulate due to its decentralized platform, and therefore some legal and governance problems have to be expected. That dependence on cryptocurrency also possibly invites the potential instability of an economy and market manipulation, too-both detrimental to its integration in the sustainable energy framework (Wenhua et al., 2023). There are another several challenges to overcome, including:

- High energy consumption: Many blockchain networks, especially those using PoW consensus mechanisms like Bitcoin, consume significant amounts of energy. This can lead to a large carbon footprint.
- Scalability issues: Blockchain networks often struggle with scalability. As the number of transactions increases, the network can become slower and less efficient, which can hinder its widespread adoption for environmental applications.
- Regulatory uncertainty: The regulatory landscape for blockchain technology is still evolving. Uncertainty around regulations can create barriers to the implementation of blockchain solutions for environmental sustainability.
- Data integration: Integrating real-world environmental data into blockchain systems can be challenging. Reliable and accurate data is crucial for the effectiveness of blockchain applications in sustainability, but obtaining and verifying this data can be difficult.
- Cost: Implementing blockchain technology can be expensive. The costs associated with developing and maintaining blockchain systems can be a barrier for many organizations, especially those in developing countries.

Despite these challenges, ongoing advancements and innovations in blockchain technology continue to address these issues, making it a promising tool for environmental sustainability in the future.

4.6.5 Latest Advancements on Blockchain Technology

- Decentralized energy markets: Blockchain allows one to create decentralized energy markets, which would enable direct trading among consumers themselves. Such peer-to-peer trade in energy may offer greater potential and efficient integrations of renewable sources like solar and wind into the network (Zhang et al., 2023).

- Energy transaction smart contracts: Smart contracts, powered by blockchain, automate energy transactions in a so transparent way with the least possible involvement of any intermediary. This further enhances the efficiency and cost of energy distribution (Goli et al., 2024).
- Smarter grid management: Blockchain could help in the development of SGs using real-time data to create an optimized approach to energy production, storage, and distribution. This minimizes energy wastage by maintaining a perfect balance between demand and supply while catering toward stabilization of the grids (Zhang et al., 2023).
- Encouragement toward renewable energy utilization: Blockchain-based mechanisms can incentivize the generation of renewable energy through the usage of token or credit systems for energy generated from renewable sources, which can then be traded or offset against energy cost, if so desired (Leong et al., 2024; Zhang et al., 2023).
- Increased transparency and trust: Because blockchain technology would offer a tamper-proof record of energy transactions, trust among participants on the energy market would increase. For example, this is useful to check the origin of renewable energy to make sure at each stage of the treatment this energy meets the sustainability standard (Leong et al., 2024).

4.7 Carbon Capture, Utilization, and Storage

CCUS first refers to the suite of technologies designed to capture carbon dioxide (CO_2) emissions either from sources such as power plants and industrial facilities or from the atmosphere directly. Later, the captured CO_2 can be stored underground or further utilized in various industrial processes (https://www.iea.org/energy-system/carbon-capture-utilisation-and-storage, Accessed on 20 July 2024). The process of this technology is as follows:

A. Capture: CO_2 is captured from emission sources using methods like post-combustion, pre-combustion, and oxy-fuel combustion. Direct air capture is also an option, though more energy-intensive.
B. Transport: The captured CO_2 is compressed and transported via pipelines, ships, rail, or road tankers to storage or utilization sites.
C. Storage/Utilization: CO_2 can be stored in deep geological formations such as depleted oil and gas reservoirs, coalbeds, or saline aquifers. Alternatively, it can be used in industrial processes like enhanced oil recovery or converted into products like chemicals and building materials.

Also, its most important uses can be mentioned below:

- Power generation: Reducing emissions from coal and gas-fired power plants.
- Industrial processes: Capturing CO_2 from cement, steel, and chemical production.
- DAC: Removing CO_2 directly from the atmosphere to address legacy emissions.

- Enhanced Oil Recovery (EOR): Injecting CO_2 into oil fields to increase oil extraction.
- Product manufacturing: Using CO_2 to create products like synthetic fuels, plastics, and concrete.

CCUS has already been deployed in multiple industrial processes, fuel transformation, and power generation in 45 commercial facilities. While CCUS deployment to date has been slower than anticipated, over the last few years momentum has really picked up, with more than 700 projects in various stages of development across the CCUS value chain. Compared with 2023, announced storage for 2030 has increased by 70%, and announced capture by 35%. This places the theoretical total captured in 2030 at around 435 million tons a year and brings announced storage to approximately 615 Mt a year. This momentum is positive, but it represents roughly only 40% and 60%, respectively, of the circa 1 Gt CO_2 per year captured and stored in NZE scenario (https://www.iea.org/energy-system/carbon-capture-utilisation-and-storage, Accessed on 20 July 2024; Mikunda et al., 2021).

By 2070, about 40% of captured CO_2 in the power sector will be linked with bioenergy. By 2070, heavy industry will capture about a quarter of CO_2, since reducing emissions is difficult or, in the case of cement, impossible. Another 30% is made up of hydrogen, ammonia, and biofuels. DAC contributes another 7%. Transport, industry, buildings, and power are all fully decarbonized as hydrogen use rises sevenfold to 520 Mt by 2070. Low-carbon hydrogen provides around 6% of cumulative emissions savings in the Sustainable Development Scenario, while fossil-based production equipped with CCUS supplies 40% of hydrogen demand. Removing carbon from the energy system offsets emissions that are technically difficult or very expensive to cut. Moreover, it may simultaneously help offset emissions outside the energy sector if the latter shows inadequate progress (https://www.iea.org/energy-system/carbon-capture-utilisation-and-storage, Accessed on 20 July 2024).

4.7.1 Methods of Carbon Capture

CCUS can capture CO_2 from many sources of various types, such as power plants, factories, or natural gas processing plants. After capture, the CO_2 is then transported and stored underground in geological formations. Different ways of carbon capture are meant for reducing the problem of global climate change; the top 10 are shown in Fig. 4.2. Their more details can be found in Madejski et al. (2022).

4.7.2 CCUS for Sustainable Energy Planning

CCUS is a versatile technology that supports the transition to a sustainable energy system by addressing several key challenges.

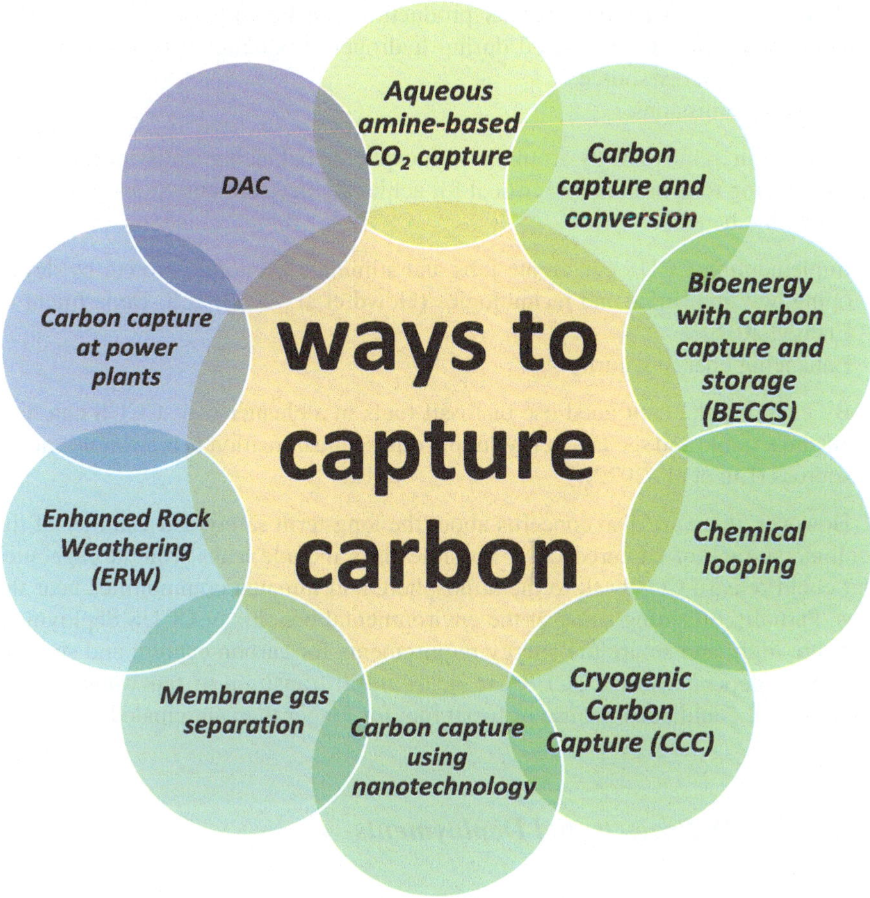

Fig. 4.2 Ten most important ways to capture carbon

- Reducing emissions

  CCUS captures the CO_2 emissions from various sources, including power plants and industrial processes, and prevents them from entering the atmosphere. This represents an important role in reducing greenhouse gas emissions to alleviate global climate change.
- Supporting hard-to-abate sectors

  Some industries, such as cement and steel production, have emissions that are difficult to eliminate through other means. CCUS provides a viable solution for these sectors, helping them reduce their carbon footprint.
- Enabling low-carbon hydrogen production

Hydrogen is a clean fuel, but its production can be carbon-intensive. CCUS can capture the CO_2 produced during hydrogen production, making it a more sustainable energy source.

- Balancing emissions

  CCUS can remove CO_2 from the atmosphere, balancing emissions that are challenging to avoid. This is critical for achieving net-zero emissions goals.

- Economic benefits

  Implementing CCUS can create jobs and stimulate economic growth by developing new industries and technologies (Gowd et al., 2023; U.S. Department of Energy, 2016).

- Enhancing energy security

  By enabling the continued use of fossil fuels in a cleaner way, CCUS can help ensure a stable and secure energy supply during the transition to renewable energy sources (Lin et al., 2022).

However, there are also concerns about the long-term safety and viability of the geologic storage of captured CO_2. Possible risks include leaks from storage sites that could release CO_2 back to the atmosphere and threaten communities near the sites. Partially offsetting some of the environmental benefits of CCUS deployment costs are high, and so are the energy requirements for carbon capture and storage. Moreover, dependence on CCUS can delay a full transition to renewable energy sources, as it could keep the use of fossil fuel for longer than anticipated.

4.7.3 CCUS Projects and Deployments

A total of 871 patents relating to CCUS technologies were registered by Shell, Chevron, BP, ExxonMobil, and Saudi Aramco between 2000 and 2022. Choosing those five firms may have been a consequence of scholarly evidence showing that only two-thirds of industrial greenhouse gas emissions over the past two centuries could be attributed to a few companies, most of which are oil and gas companies. Aside from having large market capitalization, extended scale of operations, and large production capacities, these companies are well known as leaders within their respective industries. They also have huge employee bases, significant revenues, and high net income (Truong et al., 2024). The occurrence of seven distinct CCUS technology areas developed by the five companies is shown in Table 4.1. A brief description of the seven CCUS-related technologies in the indicated fields follows. The comparison of CCUS efforts of five major oil companies is shown in Table 4.2.

I. Chemical industry: CCUS technologies in the chemical industry involve the capture of CO_2 emissions from the manufacture of chemicals production processes: wrapped up and stored, or used in the manufacture of new chemical products.

II. Capture and disposal of greenhouse gases: The principal techniques put into practice are post-combustion and pre-combustion capture, plus oxy-fuel combustion, which would separate CO_2 and other greenhouse gases from the atmosphere before emission. In this regard, after capture, the gases are stored forever or used in Enhanced Oil Recovery processes.
III. Oil refining/petrochemicals: CCUS in refining of oil and production of petrochemicals involves capturing CO_2 from such refining processes as catalytic cracking and steam reforming. Captured CO_2 can be utilized either in the production of chemicals or stored in underground repositories.
IV. Combustion technology with mitigation: Oxy-fuel combustion and chemical looping combustion are technologies to reduce CO_2 resulting from the combustion of fuels. These specifically make the fuels burn in a way to make it easier to capture the source of CO_2, hence mitigating its release.
V. Enabling technologies: These are technologies that play a supporting role in realizing CCUS. These include advanced sensors, control systems, and digital platforms. They improve monitoring, optimize efficiency in CO_2 capture, and enhance efficiency in storage with reduced operation costs.
VI. Metal processing Technologies: In the context of metals industries, CCUS captures CO_2 from production processes related to iron and steel production, recognized as one of the major CO_2-emitting sectors. Techniques providing carbon capture from blast furnaces or direct reduction processes help in combating the large carbon footprint emanating from metal production.
VII. Mineral technology processing: CCUS technologies in mineral processing industries capture CO_2 emissions captured from processes involved in cement and lime production, among others. It grants significant amounts of CO_2 emissions during calcination—a process in limestone treatment—which CCUS mitigates by capture or integration into new mineral forms.

Table 4.1 Number and percentage of patents for each technology (Truong et al., 2024)

Technology	Number and percentage of patents
Chemical industry	438 (40.4%)
Capture and disposal of greenhouse gases	270 (24.9%)
Oil refining/petrochemicals	242 (22.3%)
Combustion technology with mitigation	97 (9.0%)
Enabling technologies	17 (1.6%)
Technologies for metal processing	16 (1.5%)
Technology for processing minerals	3 (0.3%)

Table 4.2 Comparison of CCUS efforts of five major oil companies (Truong et al., 2024)

Oil and gas company	Technology focus
Shell	Invests in CCS infrastructure for industrial sites and explores CCUS deployment in a test center, as well as engaging in CCUS projects including oil sands, natural gas, refineries, and chemical plants
ExxonMobil	Researches and advances technologies for efficient CO_2 capture from industrial facilities and direct air capture (DAC) from various sources, including natural gas processing and refineries
Chevron	With regard to carbon capture at oil sands, Chevron primarily focuses on building CCS infrastructure for refineries and chemical plants. In addition, Chevron is involved in CO_2 capture technology and transportation pilot projects
BP	A variety of technologies are being developed, including CCS for gas-fired power plants, blue hydrogen production, the integration of renewable energy into industrial processes, and the development of carbon capture and storage infrastructure for industrial sources
Saudi Aramco	Using captured CO_2, Aramco enhances oil recovery rates, primarily from natural gas liquids. They develop carbon capture and storage infrastructure specifically for EOR projects

4.8 New Trends in Cutting-Edge Technologies

As renewable energy technologies grow in popularity, more people are becoming aware of the benefits of switching to cleaner solutions, such as green energy. Energy storage, electric vehicles, heat pumps, hydrogen technologies, smart electricity grids, and more alternatives to coal, oil, and gas are examples of how many industries are developing and implementing sustainable solutions (https://knowhow.distrelec.com/energy-and-power/top-10-innovative-technologies-in-sustainable-energy-sector/). Technological advances allow us to measure and monitor more things, allowing us to build a more sustainable environment for all. Here are a few of the trends we see in the energy industry going forward.

4.8.1 Solar-Powered Trains

Solar-powered trains represent a significant advancement in sustainable energy development, offering a cleaner and more efficient mode of transportation. By harnessing solar energy through photovoltaic (PV) panels installed on train carriages or along railway tracks, these trains can reduce reliance on fossil fuels and lower greenhouse gas emissions. This innovation not only contributes to the reduction of the transportation sector's carbon footprint but also promotes the use of renewable energy sources, aligning with global sustainability goals. As countries invest in solar-powered rail systems, they pave the way for a greener future, enhancing energy security and fostering environmental stewardship (Sain et al., 2020).

4.8.2 Floating Solar Farms

One innovative approach to sustainable energy development is floating solar farms. These solar power systems are installed on bodies of water, such as reservoirs, lakes, and even oceans. Floating solar farms offer several advantages: they reduce land use conflicts, improve solar panel efficiency due to the cooling effect of water, and minimize water evaporation from the bodies they cover. Additionally, they can be integrated with existing hydropower plants to create hybrid systems, enhancing overall energy production and reliability. This technology represents a promising solution for maximizing renewable energy generation while addressing land and water resource challenges (Ogunjo et al., 2023).

4.8.3 Food Waste Solar Panels

Food waste solar panels, such as the AuREUS system, are revolutionizing sustainable energy development by converting agricultural waste into renewable energy. This innovative technology uses luminescent particles derived from fruit and vegetable waste to capture ultraviolet (UV) light and convert it into visible light, which is then transformed into electricity by PV cells. Unlike traditional solar panels, AuREUS can generate power even on cloudy days, significantly enhancing energy efficiency and reliability. Upcycling food waste not only reduces environmental impact but also supports circular economy principles, making it a promising solution for a greener future (Sridhar et al., 2021).

4.8.4 Bladeless Wind Energy

Bladeless wind energy is an innovative approach to sustainable energy development, offering several advantages over traditional wind turbines. These turbines, which utilize oscillation to generate power, eliminate the need for large, rotating blades, thereby reducing noise pollution and the risk to wildlife, particularly birds. Additionally, bladeless turbines are more compact and can be easily integrated into urban environments, making them suitable for a wider range of locations. Their simpler design also means lower maintenance costs and a smaller environmental footprint, contributing to a more sustainable and efficient renewable energy solution (Elsayed et al., 2024).

4.8.5 Biometric Authentication Approach

Biometric authentication, while primarily known for enhancing security and convenience in digital transactions, also holds potential as an innovative approach to sustainable energy development. By integrating biometric systems into energy management, we can create more secure and efficient ways to control access to energy resources and infrastructure. For instance, biometric authentication can ensure that only authorized personnel can access critical energy facilities, reducing the risk of unauthorized usage and enhancing overall energy security. Additionally, this technology can streamline the monitoring and management of energy consumption in SGs and buildings, leading to more efficient energy use and reduced waste (Salturk & Kahraman, 2024).

4.8.6 Rechargeable Tires

Rechargeable tires represent a groundbreaking innovation in sustainable energy development, particularly for EVs. These tires are designed to harness and store energy generated during vehicle operation, which can then be used to extend the range of EVs and improve overall energy efficiency. By integrating advanced material science and sustainable automotive technologies, rechargeable tires not only meet stringent environmental regulations but also cater to the growing consumer demand for greener transportation solutions. This innovation holds the potential to significantly reduce the carbon footprint of the automotive industry, contributing to a more sustainable future (Rechargeable Tyres Market Size, Share, Growth | CAGR of 4.4%, 2024).

4.8.7 3D-Printed Solar Energy Trees

3D-printed solar energy trees are an innovative technology that merges advanced 3D printing with renewable energy solutions to create sustainable energy sources. Developed by researchers at VTT Technical Research Centre of Finland, these "trees" feature flexible solar panels as leaves, which can harvest solar energy, wind energy, and temperature fluctuations. The trunks are made from wood-based biocomposites, making them both eco-friendly and mass-producible. This technology not only provides a visually appealing way to generate clean energy but also offers a versatile solution for powering small devices and contributing to the reduction of carbon emissions. By mimicking natural processes, these solar energy trees represent a significant step forward in sustainable energy development (Joshna & Dudhe, 2022; Li et al., 2023).

4.8.8 Carbon Nanotubes (CNTs)

CNTs are a revolutionary technology in the realm of sustainable energy development. These cylindrical nanostructures, composed of carbon atoms arranged in a hexagonal lattice, exhibit exceptional mechanical strength, high electrical conductivity, and thermal stability. Their unique properties make them ideal for enhancing the performance of energy storage and conversion technologies, such as batteries and supercapacitors. Additionally, CNTs can be used to create lightweight, efficient solar cells and improve hydrogen fuel cells, contributing to the reduction of greenhouse gas emissions. By integrating CNTs into various applications, we can significantly advance the development of sustainable and carbon–neutral energy solutions (Zhang et al., 2024).

4.9 Conclusion

This book chapter employs a comprehensive methodology for data collection and classification, with a particular emphasis on the pivotal objectives of sustainable energy development. The chapter commences with an examination of the most essential objectives, including economic and social goals, which are fundamental to the advancement of sustainable energy solutions. Subsequently, the chapter investigates the impact of pioneering technologies on sustainable energy development. These technologies, distinguished by their advanced, innovative, and pioneering nature, represent the cutting edge of progress in the energy sector. They integrate the latest scientific and engineering advancements, playing a pivotal role in shaping the future of sustainable energy systems. By examining the nexus of technology and sustainability, the chapter offers insights into how these developments can contribute to achieving long-term energy sustainability goals. For this purpose and among the cutting-edge technologies, here five technologies including green energy storages, SGs, biofuels, blockchain, CCUS, etc., were analyzed and investigated. About each of them, their economic and social advantages and challenges were mentioned separately, and their types were categorized. The most important type of green storages mentioned in this book chapter is the hydrogen storage device, which is very popular and trendy today. Also, biofuels and CCUS discussed here play an important role in achieving sustainable energy development goals. Regarding SGs, it can be said that the role of AI is very important in them. This issue was investigated, and the role and importance of AI in these networks were analyzed both from the point of view of consumers and suppliers. It is important to note that these technologies always have challenges, the most important of which have been briefly addressed. In the end, the new trends of cutting-edge technologies that can be effective in helping to develop sustainable energy were reviewed and examples of these new trends were introduced.

References

Alonso, A. P. (2024). Pumped hydro energy storage systems for a sustainable energy planning. In *Sustainable Energy Planning in Smart Grids* (pp. 71–89). Elsevier.

Amiri, M. M., Ameli, M. T., Aghamohammadi, M. R., Bashooki, E., Ameli, H., & Strbac, G. (2024a). Day-ahead coordination for flexibility enhancement in hydrogen-based energy hubs in presence of EVs, storage, and integrated demand response. *IEEE Access*.

Amiri, M. M., Ameli, M. T., Strbac, G., Pudjianto, D., & Ameli, H. (2024b). The role of flexibility in the integrated operation of low-carbon gas and electricity systems: A review. *Energies, 17*(9), 2187.

Arumugham, V., Ghanimi, H. M., Pustokhin, D. A., Pustokhina, I. V., Ponnam, V. S., Alharbi, M., Krishnamoorthy, P., & Sengan, S. (2023). An artificial-intelligence-based renewable energy prediction program for demand-side management in smart grids. *Sustainability, 15*(6), 5453.

Aydin, M. I., & Dincer, I. (2022). An assessment study on various clean hydrogen production methods. *Energy, 245*, 123090.

Aydin, M. I., Karaca, A. E., Qureshy, A. M., & Dincer, I. (2021). A comparative review on clean hydrogen production from wastewaters. *Journal of Environmental Management, 279*, 111793.

Biofuels: a sustainable solution for a green energy future. https://www.spectrafuels.com/biofuels-a-sustainable-solution-for-a-green-energy-future/

Cheng, F., Luo, H., & Colosi, L. M. (2020). Slow pyrolysis as a platform for negative emissions technology: An integration of machine learning models, life cycle assessment, and economic analysis. *Energy Conversion and Management, 223*, 113258.

Cui, M. L., Feng, T. T., & Wang, H. R. (2023). How can blockchain be integrated into renewable energy?—A bibliometric-based analysis. *Energy Strategy Reviews, 50*, 101207.

Demırbas, A. (2017). The social, economic, and environmental importance of biofuels in the future. *Energy Sources, Part b: Economics, Planning, and Policy, 12*(1), 47–55.

Dincer, I., & Aydin, M. I. (2023). New paradigms in sustainable energy systems with hydrogen. *Energy Conversion and Management, 283*, 116950.

Effah, D., Chunguang, B., Appiah, F., Agbley, B. L. Y., & Quayson, M. (2021, December). Carbon emission monitoring and credit trading: the blockchain and IOT approach. In *2021 18th International Computer Conference on Wavelet Active Media Technology and Information Processing (ICCWAMTIP)* (pp. 106–109). IEEE.

Elalfy, D. A., Gouda, E., Kotb, M. F., Bureš, V., & Sedhom, B. E. (2024). Comprehensive review of energy storage systems technologies, objectives, challenges, and future trends. *Energy Strategy Reviews, 54*, 101482.

Elsayed, A. M., Gaheen, O. A., Abdelrahman, M. A., & Aziz, M. A. (2024). An experimental investigation of a solar chimney integrated with a bladeless wind turbine for sustainable energy harvesting. *Energy*, 132154.

Goli, A., Bozanic, D., Zhou, F., & Ali, I. (2024). Editorial: The role of blockchain technology toward a sustainable energy future. *Frontiers in Energy Research, 12*, 1361080. https://doi.org/10.3389/fenrg.2024.1361080

Gowd, S. C., Ganeshan, P., Vigneswaran, V. S., Hossain, M. S., Kumar, D., Rajendran, K., Ngo, H. H., & Pugazhendhi, A. (2023). Economic perspectives and policy insights on carbon capture, storage, and utilization for sustainable development. *Science of the Total Environment, 883*, 163656.

Gunnarsdóttir, I., Davidsdottir, B., Worrell, E., & Sigurgeirsdóttir, S. (2021). Sustainable energy development: History of the concept and emerging themes. *Renewable and Sustainable Energy Reviews, 141*, 110770.

Hassan, Q., Sameen, A. Z., Salman, H. M., Jaszczur, M., & Al-Jiboory, A. K. (2023). Hydrogen energy future: Advancements in storage technologies and implications for sustainability. *Journal of Energy Storage, 72*, 108404.

Herc, L., Pfeifer, A., Duić, N., & Wang, F. (2022). Economic viability of flexibility options for smart energy systems with high penetration of renewable energy. *Energy, 252*, 123739.

https://knowhow.distrelec.com/energy-and-power/top-10-innovative-technologies-in-sustainable-energy-sector/

https://www.iea.org/energy-system/carbon-capture-utilisation-and-storage, Accessed on 20 July 2024

Hydrogen storage: A challenge for the development of the industry. https://www.terega.fr/en/our-activities/hydrogen/hydrogen-storage-a-challenge-for-the-development-of-the-industry/

Irham, A., Roslan, M. F., Jern, K. P., Hannan, M. A., & Mahlia, T. I. (2024). Hydrogen energy storage integrated grid: A bibliometric analysis for sustainable energy production. *International Journal of Hydrogen Energy, 63*, 1044–1087.

Jangir, M., Rawat, N. S., & Kumar, H. (2024). Solid-state materials for hydrogen storage. In *Challenges and opportunities in green hydrogen production* (pp. 491–514). Springer Nature Singapore.

Joshna, A., & Dudhe, R. (2022). A case study and review on current trends in solar energy: Path to a greener environment. In *Proceedings of First International Conference on Computational Electronics for Wireless Communications: ICCWC 2021* (pp. 139–153). Springer Singapore.

Kabeyi, M. J. B., & Olanrewaju, O. A. (2022, January). The use of smart grids in the energy transition. In *2022 30th Southern African Universities Power Engineering Conference (SAUPEC)* (pp. 1–8). IEEE.

Khalid, M. (2024). Smart grids and renewable energy systems: Perspectives and grid integration challenges. *Energy Strategy Reviews, 51*, 101299.

Leong, W. Y., Leong, Y. Z., & San Leong, W. (2024, July). Blockchain technology in next generation energy management system. In *2024 7th International Conference on Green Technology and Sustainable Development (GTSD)* (pp. 15–19). IEEE.

Li, J., Li, M., Koh, J. J., Wang, J., & Lyu, Z. (2023). 3D-printed biomimetic structures for energy and environmental applications. *DeCarbon*, 100026.

Li, X., & Palazzolo, A. (2022). A review of flywheel energy storage systems: State of the art and opportunities. *Journal of Energy Storage, 46*, 103576.

Lin, Q., Zhang, X., Wang, T., Zheng, C., & Gao, X. (2022). Technical perspective of carbon capture, utilization, and storage. *Engineering, 14*, 27–32.

Liu, Y., Cruz-Morales, P., Zargar, A., Belcher, M. S., Pang, B., Englund, E., Dan, Q., Yin, K., & Keasling, J. D. (2021). Biofuels for a sustainable future. *Cell, 184*(6), 1636–1647.

Madejski, P., Chmiel, K., Subramanian, N., & Kuś, T. (2022). Methods and techniques for CO_2 capture: Review of potential solutions and applications in modern energy technologies. *Energies, 15*(3), 887.

Malik, K., Capareda, S. C., Kamboj, B. R., Malik, S., Singh, K., Arya, S., & Bishnoi, D. K. (2024). Biofuels production: A review on sustainable alternatives to traditional fuels and energy sources. *Fuels, 5*(2), 157–175.

Meydani, A., Shahinzadeh, H., Ramezani, A., Moazzami, M., Nafisi, H., & Askarian-Abyaneh, H. (2024, February). Comprehensive Review of Artificial Intelligence Applications in Smart Grid Operations. In *2024 9th International Conference on Technology and Energy Management (ICTEM)* (pp. 1–13). IEEE.

Mikunda, T., Brunner, L., Skylogianni, E., Monteiro, J., Rycroft, L., & Kemper, J. (2021). Carbon capture and storage and the sustainable development goals. *International Journal of Greenhouse Gas Control, 108*, 103318.

Moreno Escobar, J. J., Morales Matamoros, O., Tejeida Padilla, R., Lina Reyes, I., & Quintana Espinosa, H. (2021). A comprehensive review on smart grids: Challenges and opportunities. *Sensors, 21*(21), 6978.

Mostafa, K., & Saad, A. (2024). Optimizing decentralized systems with multimodal AI: advanced strategies for enhancing performance, scalability, and real-time decision-making in distributed architectures. *Applied Research in Artificial Intelligence and Cloud Computing, 7*(6), 135–160.

Muhammed, N. S., Gbadamosi, A. O., Epelle, E. I., Abdulrasheed, A. A., Haq, B., Patil, S., Al-Shehri, D., & Kamal, M. S. (2023). Hydrogen production, transportation, utilization, and storage: Recent advances towards sustainable energy. *Journal of Energy Storage, 73*, 109207.

Nastasi, B., Markovska, N., Puksec, T., Duić, N., & Foley, A. (2022). Renewable and sustainable energy challenges to face for the achievement of sustainable development goals. *Renewable and Sustainable Energy Reviews, 157*, 112071.

Nath, S. (2024). Biotechnology and biofuels: Paving the way towards a sustainable and equitable energy for the future. *Discover Energy, 4*(1), 8.

Ogunjo, S., Olusola, A., & Olusegun, C. (2023). Potential of using floating solar photovoltaic and wind farms for sustainable energy generation in an existing hydropower station in Nigeria. *Clean Technologies and Environmental Policy, 25*(6), 1921–1934.

Prasad, S., Yadav, K. K., Kumar, S., Pandita, P., Bhutto, J. K., Alreshidi, M. A., Ravindran, B., Yaseen, Z. M., Osman, S. M., & Cabral-Pinto, M. M. (2024). Review on biofuel production: Sustainable development scenario, environment, and climate change perspectives—A sustainable approach. *Journal of Environmental Chemical Engineering, 12*(2), 111996.

Rabi, A. M., Radulovic, J., & Buick, J. M. (2023). Comprehensive review of compressed air energy storage (CAES) technologies. *Thermo, 3*(1), 104–126.

Rampai, M. M., Mtshali, C. B., Seroka, N. S., & Khotseng, L. (2024). Hydrogen production, storage, and transportation: Recent advances. *RSC Advances, 14*(10), 6699–6718.

Rechargeable Tyres Market Size, Share, Growth | CAGR of 4.4%. (2024). https://market.us/report/rechargeable-tyres-market

Rejeb, A., Rejeb, K., Zrelli, I., Süle, E., & Iranmanesh, M. (2024). Blockchain technology in the renewable energy sector: A co-word analysis of academic discourse. *Heliyon, 10*(8).

Ren, L., Li, Y., Zhang, N., Li, Z., Lin, X., Zhu, W., Lu, C., Ding, W., & Zou, J. (2023). Nanostructuring of Mg-based hydrogen storage materials: recent advances for promoting key applications. *Nano-Micro Letters, 15*(1), 93.

Sadiqa, A. (2023). *Sustainable energy transition for Pakistan: Assessing the role energy, water supply, social and gender equity dimensions.*

Sain, C., Banerjee, A., Biswas, P. K., & Padmanaban, S. (2020). A state-of-the-art review on solar-powered energy-efficient PMSM drive smart electric vehicle for sustainable development. *Advances in Greener Energy Technologies*, 231–258.

Salturk, S., & Kahraman, N. (2024). Deep learning-powered multimodal biometric authentication: integrating dynamic signatures and facial data for enhanced online security. *Neural Computing and Applications*, 1–12.

Saritha, K., Sharma, S., Dutt, A., Shrivastava, A., Rao, A. K., Jawad, A. H., & Saikumar, A. (2024). Green energy storage solutions: A research. In *E3S web of conferences* (Vol. 552, p. 01129). EDP Sciences.

Shah, I. A., Jhanjhi, N. Z., Amsaad, F., & Razaque, A. (2022). The role of cutting-edge technologies in industry 4.0. In *Cyber Security Applications for Industry 4.0* (pp. 97–109). Chapman and Hall/CRC.

Sharipov, K., Abdurashidova, N., Valiyeva, A., Tuychieva, V., Kholmatova, M., & Minarova, M. (2022, December). A systematic mapping study of using the cutting-edge technologies in marketing: the state of the art of four key new-age technologies. In *International Conference on Next Generation Wired/Wireless Networking* (pp. 381–389). Springer Nature Switzerland.

Sridhar, A., Kapoor, A., Kumar, P. S., Ponnuchamy, M., Balasubramanian, S., & Prabhakar, S. (2021). Conversion of food waste to energy: A focus on sustainability and life cycle assessment. *Fuel, 302*, 121069.

Sundaram, T., Rajendran, S., Gnanasekaran, L., Rachmadona, N., Jiang, J. J., Khoo, K. S., & Show, P. L. (2023). Bioengineering strategies of microalgae biomass for biofuel production: Recent advancement and insight. *Bioengineered, 14*(1), 2252228.

Taherdoost, H. (2024). Blockchain Integration and Its Impact on Renewable Energy. *Computers, 13*(4), 107.

Truong, T. H., Lin, B. W., Lo, C. H., Tung, C. P., & Chao, C. W. (2024). Possible pathways for low carbon transitions: Investigating the efforts of oil companies in CCUS technologies. *Energy Strategy Reviews, 54*, 101421.

U.S. Department of Energy (2016). *Carbon capture, utilization and storage: Climate change, economic competitiveness, and energy security.*

Wenhua, Z., Qamar, F., Abdali, T. A. N., Hassan, R., Jafri, S. T. A., & Nguyen, Q. N. (2023). Blockchain technology: Security issues, healthcare applications, challenges and future trends. *Electronics, 12*(3), 546.

Yu, Q., Wang, Q., Tan, X., Fang, G., & Meng, J. (2019). A review of compressed-air energy storage. *Journal of Renewable and Sustainable Energy, 11*(4).

Zhang, C., Zhu, B., & Liu, Z. (2023). The potential of blockchain technology in advancing sustainable energy: A study on the mediating role of specialization in the growth of Chinese hidden champions. *Frontiers in Energy Research, 11*, 1272942.

Zhang, X., Han, G., & Zhu, S. (2024). Flash nitrogen-doped carbon nanotubes for energy storage and conversion. *Small (Weinheim an der Bergstrasse, Germany), 20*(3), 2305406.

Zibaeirad, A., Koleini, F., Bi, S., Hou, T., & Wang, T. (2024). *A comprehensive survey on the security of smart grid: Challenges, mitigations, and future research opportunities.* arXiv preprint arXiv: 2407.07966.

Chapter 5
On the Essence of Coordination for Sustainable Integrated Power and Natural Gas Systems: Basics, Progress, and Challenges

Mostafa Shabanian-Poodeh and Miadreza Shafie-khah

Abstract Energy infrastructures from different sectors, such as power networks and natural gas networks, are traditionally planned and operated dependently, necessitating sufficient coordination. The lack of coordination results in low energy efficiency and high operating costs. Meanwhile, market competitions and natural disasters might disrupt the coordination of such energy system. In this chapter, the basics and challenges for implementing integrated power and gas systems (IPGSs) are presented. Additionally, recent studies that have proposed methods to maintain the coordination and safe operation of the IPGS in energy market environment and mitigate catastrophic events damage are compared and analyzed. Moreover, the base operation problem of IPGS is presented, which is commonly used in the literature. Additionally, convexification methods including second-order cone relaxation and piecewise linearization methods are explained. Overall, this chapter presents an overview on the progress and challenges for implementing IPGS in market and enhancing its resilience. Finally, the basic optimization problem for the operation of IPGS is implemented in GAMS, and the numerical results with a market perspective are presented to effectively demonstrate the interdependencies of such energy systems which highlights the importance of coordination.

Keywords Resilient energy systems · Energy market · Energy systems coordination · Optimization · Convexification

M. Shabanian-Poodeh (✉)
Department of Electrical Engineering, Faculty of Engineering, University of Isfahan, Isfahan, Iran
e-mail: mostafashabanian188@gmail.com

M. Shafie-khah
Research and Innovation Division, Nowocert, Dublin, Ireland

I. Rahimi et al. (eds.), *Sustainable Supply Chain of Renewable Energy Networks*, Energy Systems in Electrical Engineering, https://doi.org/10.1007/978-981-96-4940-2_5

Nomenclatures

Indices

b, d	Index of the bus
c	Index of the conventional generators
e	Index of the P2G units
gf	Index of the G2P units
n, m	Index of the gas nodes

Parameters and Constants

B_l	Susceptance of transmission line
C_c^{conv}	Operation cost of conventional generation unit c (\$/MW)
C_w^{Well}	Operation cost of gas well w (\$/Mm^3)
C^{LS}	Cost of load shedding (\$)
$F_{e,\min}^{\text{P2G}}/F_{e,\max}^{\text{P2G}}$	Minimum/maximum gas production of P2G e (Mm^3)
$F_{gf,\min}^{\text{G2P}}/F_{gf,\max}^{\text{G2P}}$	Minimum/maximum gas consumption of G2P gf (Mm^3)
$\overline{F_w^{\text{Well}}}/\underline{F_w^{\text{Well}}}$	Maximum/minimum gas production of gas well w (Mm^3)
$F_p^{\max}$	Maximum gas flow on pipeline p (Mm^3)
F_n^L	Natural gas load in gas node n (Mm^3)
GHV	Calorific Value of Natural Gas (MJ/m^3)
$P_c^{\min}/P_c^{\max}$	Minimum/maximum power generation of conventional generator c (MW)
P_b^L	Electric load (MW)
$\overline{\text{PF}_l}$	Maximum power flow on line l (MW)
$\rho_n^{\min}/\rho_n^{\max}$	Minimum/maximum pressure in gas node n (bar)
γ	Maximum gas pressure boost ratio of the compressor
μ_e^{P2G}	Conversion coefficient of P2G unit e
μ_{gf}^{G2P}	Conversion coefficient of G2P unit gf

Variables

F_w^{Well}	Natural gas generation of well w (Mm^3)
F_e^{P2G}	Natural gas generation of P2G e (Mm^3)
F_{gf}^{G2P}	Natural gas consumption of G2P gf (Mm^3)
F_p	Natural gas flow on pipeline p (Mm^3)
F_n^{LS}	Amount of natural gas load shedding at gas node n (Mm^3)
P_e^{P2G}	Power consumption of P2G e (MW)

P_{gf}^{G2P}	Power generation of G2P *gf* (MW)
P_c	Power generation of conventional generation unit *c* (MW)
P_b^{LS}	Amount of power load shedding at bus *b* (MW)
PF_l	Power flow on transmission line *l* (MW)
Z	Objective function value ($)
θ_b	Voltage angle at bus *b* (rad)

5.1 Introduction

Integrated power-gas systems (IPGSs) represent an innovative approach to energy infrastructure, seamlessly combining power and gas energy systems to establish a resilient and sustainable energy ecosystem (Hemmati et al., 2021). At the heart of IPGSs are the transformative processes of power-to-gas (P2G) and gas-to-power (G2P), enabling the conversion of excess renewable energy into storable gases and their subsequent reconversion into electricity as required (Gharai et al., 2023). This mutually beneficial relationship between power and gas networks is coordinated by a central coordinator, ensuring efficient resource utilization and dynamic balancing of supply and demand (Armioun et al., 2023). The framework of this concept is illustrated in Fig. 5.1.

The optimal operation of IPGS must be presented as a mathematical optimization model that encompasses the crucial limitations and characteristics of these systems. Solving this model allows for the optimal scheduling of generation units to achieve specific goals, typically minimizing operation costs (Zandrazavi et al., 2023). However, among the system constraints, some are often non-convex, introducing significant computational challenges (Kocuk et al., 2016). To address these challenges, convexification methods such as piecewise linearization (PWL) and second-order cone relaxation (SOCR) are commonly used. These techniques make the model more tractable and applicable to real-world problems, often categorized as large-scale

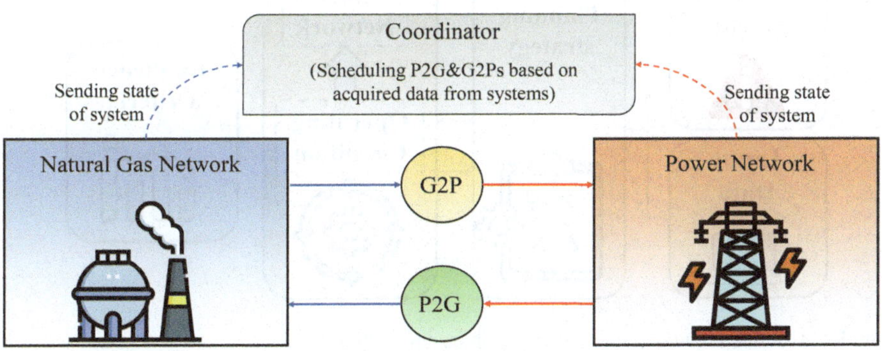

Fig. 5.1 Structure of the IPGS and its coordination scheme

planning and operation problems. Planning for IPGS can be aimed at enhancing the resilience and sustainability of the energy supply, especially in the face of natural disasters and papers that propose operation problems often aim to solve market participation issues. By integrating power and gas energy systems, IPGS expands across large geographic areas, yet this expansion also renders them more vulnerable to natural events such as earthquakes, floods, and other environmental challenges (Gasser et al., 2021). Despite their numerous advantages, such as increased efficiency (Daneshvar et al., 2023) and reduced emissions (Li, 2022), IPGSs also face heightened risks due to their extended reach, necessitating strategic planning for infrastructure to mitigate potential disruptions. Furthermore, the operation of IPGS plays a crucial role in enabling them to withstand such shocks (Gasser et al., 2021).

According to Fig. 5.2, the process for enhancing the resilience of begins with identifying the type of extreme event specific to a region. This is followed by determining an appropriate strategy or planning type to bolster resilience. Next, the planning and operation phases must interact, with the operator assessing and modifying plans as needed to ensure feasibility. Finally, defining resilience criteria is essential to evaluate the impact of these plans on system resilience. This approach ensures IPGS are robust, adaptable, and capable of maintaining continuous energy supply despite disruptions.

This chapter follows a structured exploration of IPGS planning and operation modeling and resilience enhancement strategies. It began with an introduction outlining the significance of IPGS and the challenges posed by non-convex optimization problems. The subsequent sections delve into the resilience of IPGS, elucidating its dimensions, planning strategies, and criteria for assessment in recent papers. Section 5.3 discusses the market of the IPGS and a comprehensive review on the recent studies to further clarify the market mechanism of IPGS. Section 5.4 starts with a detailed and basic presentation of the mathematical models for gas and power systems operation within IPGS. The chapter then delves into model convexification techniques (Sect. 5.5), specifically focusing on piecewise linearization (PWL) and

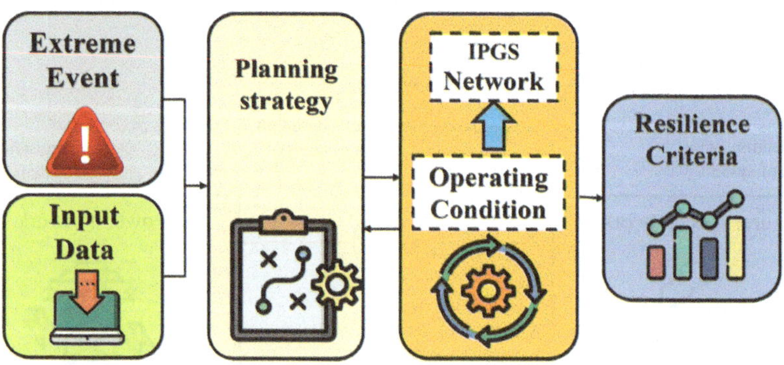

Fig. 5.2 Common roadmap for resilience enhancement studies

second-order cone relaxation (SOCR) methods to address non-convexities. Additionally, the numerical results of the operation model are demonstrated in Sect. 5.6 to effectively show the interdependence of IPGS. Section 5.7 delves into the challenges and limitations of the IPGS implementation in practice and the advancements to tackle such issues. Finally, the chapter concludes by summarizing key insights and emphasizing the importance of optimization techniques in ensuring the efficiency, reliability, and resilience of IPGS, while also suggesting potential avenues for future research.

5.2 Resilience of IPGS

The resilience of IPGS refers to the capacity of these systems to anticipate, absorb, adapt to, and recover from disruptions, ensuring a continuous and reliable energy supply. Resilience in this context encompasses several dimensions: robustness, the ability to withstand initial shocks; redundancy, having backup systems in place; resourcefulness, the ability to adapt and reconfigure resources in response to disruptions; and rapid recovery, the ability to quickly return to normal operations (Xu et al., 2023). Strategic planning plays a critical role in enhancing the resilience of IPGS. This involves the careful design and processes of coupling devices, leveraging advanced technologies to optimize resource use and balance supply and demand dynamically. It also requires coordinated response strategies to manage and mitigate the impacts of natural hazards and extreme events.

Moreover, defining clear resilience criteria is essential for this process. These criteria provide measurable benchmarks for assessing and improving the resilience of IPGS. They can include metrics such as system reliability, recovery time, adaptability, and robustness. By establishing and adhering to these criteria, stakeholders can systematically identify vulnerabilities, implement effective countermeasures, and ensure the ongoing stability and functionality of IPGS.

5.2.1 Planning Strategies for Resilience Enhancement of IPGS

Enhancing the resilience of IPGS through planning falls under long-term measures, which involve incorporating redundancy and fortifying the system, commonly referred to as expansion planning. The process of adding redundancy inherently strengthens the system's ability to withstand severe incidents, allowing it to sustain a higher load throughout the event. Hardening also includes fortifying vulnerable components such as pipes and transmission lines. However, in specific scenarios, these measures may be implemented individually or in various combinations. For example, in references (Hasanzad & Rastegar, 2022; He et al., 2018), the focus is

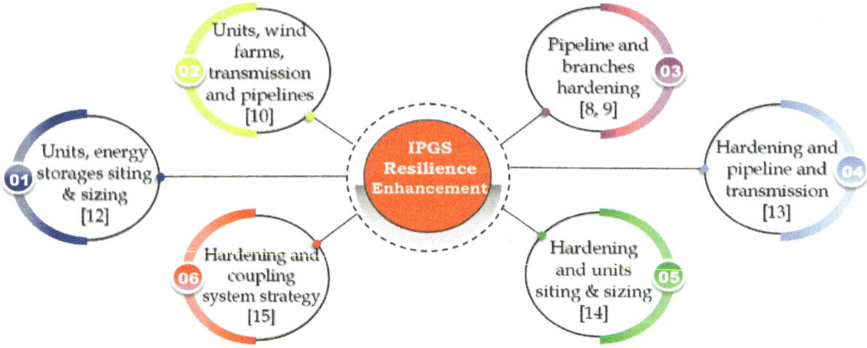

Fig. 5.3 Long-term measures to improve the resilience of IPGS in the literature

primarily on reinforcing pipelines and transmission lines, as these measures were deemed sufficient for the outlined conditions. Conversely, expansion planning and redundancy incorporation encompass a broader spectrum, involving a larger array of resources and equipment for implementation planning. References (Aldarajee et al., 2020; Hasanzad & Rastegar, 2022; He et al., 2018) primarily focus on methods related to network development planning. Nonetheless, similar to short-term measures, there is potential for synergistic combinations of these strategies, often leading to superior outcomes. References (Li et al., 2023; Wang et al., 2022; Zhang & Ding, 2023) have investigated the impact of long-term measures on enhancing IPGS resilience, offering models that integrate various long-term measures. Moreover, Reference (Shabanian-Poodeh et al., 2024b) introduced an optimal integration method along with decentralized operation to enhance the resilience of IPGS.

Figure 5.3 provides a comprehensive overview of previous research articles introducing methodologies within the domain of long-term measures, alongside their respective findings.

5.2.2 *Resilience Criteria*

IPGS resilience criteria are essential for maintaining a consistent and reliable energy supply, given the planning decisions. These benchmarks assess a system's ability to handle disruptions and offer guidelines for improving its reliability and adaptability. In a fast-changing technological environment and amid uncertainties, these criteria are vital for developing robust systems that can endure and recover from unexpected events (Wang et al., 2022; Xie et al., 2022).

Generally, quantitative assessment of a system's resilience to severe natural disasters is divided into two main categories: operational criteria and structural criteria.

- **Operational Criteria**: These criteria are essential for assessing the IPGS's capability to maintain optimal performance during and after disturbances, making them critical for resilience planning and evaluation (Xie et al., 2022).
- **Structural Criteria**: As environmental concerns grow, criteria that evaluate the system's ability to handle extreme events, resource availability, and carbon emissions become increasingly important. These criteria align IPGS with sustainability goals and ensure its adaptability to evolving environmental conditions (Jufri et al., 2019).

Considering a combination of criteria tailored to specific characteristics and objectives provides a more holistic understanding of the system's resilience. This comprehensive approach aids in making informed decisions for improving the system's resilience to disturbances and its ability to recover efficiently.

5.3 Market of IPGS

The IPGS market has gained significant attention in recent years due to the increasing focus on renewable energy integration and the need for flexible, reliable energy infrastructure. Within this market electricity and natural gas networks are integrated with a view to improving system efficiency, reliability, and flexibility (Shabanian-Poodeh et al., 2024a). The integration also provides better attempts at harnessing renewable energy that is intermittent, better energy storage and better resource utilization (Jia et al., 2023). Recent advancements in cyber-physical interactions and distributed energy resources have further enhanced the potential of IPGS, enabling more sophisticated control strategies and optimized operations. Similarly, with the increasing pressure on the energy sector to embrace low carbon energy, the integrated power and gas systems market is expected to be a game changer in the energy balance between supply and demand market participation phenomenon (Chen et al., 2023; Hu et al., 2024; Li et al., 2023).

Optimal participation in IPGS markets is crucial for stakeholders to maximize their economic benefits while contributing to system stability and efficiency, as it enables strategic decision-making in resource allocation, pricing, and risk management across both energy sectors. In this vein, Reference (Nasiri et al., 2023) introduced a robust bi-level approach for optimal participation of multi-energy service providers. In this study, the unite commitment consideration aligned with demand response programs and energy storages has been implemented to effectively minimize the overall costs. Reference (Dimitriadis et al., 2023a) proposed a bi-level carbon trading method in the multi-energy market with high penetration of renewable energy resources. In this method, the lower-level problem deals with clearing the day-ahead electricity and natural gas markets sequentially while taking carbon emission provisioning into consideration. The upper-level problem seeks to maximize the profitability of the strategic player. Moreover, Reference (Dimitriadis et al., 2023b) investigated the effect of the power system congestion effect on the locational marginal prices and social welfare in the IPGS market using a two-stage linear model. Reference (Moradi et al., 2024) proposed a framework for participation of the energy smart hubs in local IPGS market. In this model, the profit of smart hubs are maximized and it is solved using a mathematical program with equilibrium constraints (MPEC) approach.

5.4 IPGS Operation Model

As discussed in the previous sections, the IPGS encompasses two systems: power and natural gas. This section presents the mathematical models of the crucial components within each system, which are essential for creating a comprehensive model for optimal and coordinated operation of IPGS. It is noteworthy to mention that in this chapter, the model of the power transmission system is presented.

5.4.1 Natural Gas Systems Formulation

Natural gas systems consist of many components that are crucial for the stability and safe operation of such energy systems. The natural gas system includes six main components (Faramawy et al., 2016): exploration and extraction, where geologists identify and drill for gas reserves; processing, which involves separating and treating the raw gas to remove impurities and extract natural gas liquids; transportation, using extensive pipeline networks and liquid natural gas (LNG) tankers for long-distance shipping; storage, which balances supply and demand through underground reservoirs or LNG storage facilities; distribution, managed by local distribution companies delivering gas to consumers via smaller pipelines; and consumption, where natural gas is utilized for residential heating, cooking, industrial processes, and electricity generation. Each stage ensures the safe, efficient, and reliable delivery of natural gas from production to end use.

However, including the entire process of delivering gas from generation points to consumption points would complicate the optimization problem, known as the operation problem. Therefore, simplifications are necessary to reduce complexity and expedite the solution process, maintaining a balance between accuracy and computational efficiency. In this regard, the operation model of natural gas systems typically includes simplified mathematical models of gas wells, compressors, pipelines (gas flow), and an essential component: nodal gas balance that are explained in the following subsections.

- **Gas wells**

A gas well is a drilled hole in the ground designed to extract natural gas from underground reservoirs. This process involves drilling into rock formations that contain natural gas, installing steel pipes (casings) to keep the well open, and sometimes using hydraulic fracturing to enhance gas flow (Kidnay et al., 2019). The extracted natural gas is then brought to the surface for processing and use. Ultimately, the gas is injected into the natural gas system as a source, although these sources are commonly referred to as gas wells in the literature.

A gas well has an upper bound and lower bound for its operation to ensure safety, efficiency, and economic viability. The upper bound limits the maximum flow of injected gas to prevent damage to the well infrastructure, avoid blowouts, and manage reservoir pressure for sustained production. The lower bound ensures that the pressure and flow remain sufficient to justify the cost of operation and maintain the economic feasibility of the well (Kidnay et al., 2019). These bounds help optimize production, protect the environment, and ensure the long-term viability of the gas resource. To mathematically model the abovementioned matter, inequality (5.1) can be used.

$$\underline{F_{w}^{\text{well}}} \leq F_{n,w}^{\text{well}} \leq \overline{F_{w}^{\text{well}}} \quad \forall w, n \tag{5.1}$$

where $F_{n,w}^{\text{well}}$ is the amount of injected gas by well w and node n. $\underline{F_{w}^{\text{well}}}$ and $\overline{F_{w}^{\text{well}}}$ represent minimum and maximum capacity of well w, respectively.

- **Compressors**

Compressors in the natural gas system are machines that increase the pressure of the natural gas to move it through pipelines. They play a vital role in ensuring the continuous flow of gas from production locations to processing plants, storage facilities, and final consumers (Kostowski et al., 2015). Compressors ensure that the gas travels efficiently over long distances and through various stages of the natural gas system. In the context of simplified model of compressors, the nodes that contain compressors between them, will have the following relation (Li et al., 2022):

$$\rho_n^2 \leq \gamma^2 \rho_m^2 \quad \forall m, n \tag{5.2}$$

where ρ_n is the pressure at node n and γ is the maximum gas pressure boost ratio of the compressor. Moreover, the pressure of the nodes in the natural gas system must be in a certain bound for the safety which is guaranteed by (5.3).

$$\rho_n^{\min} \leq \rho_n \leq \rho_n^{\max} \quad \forall m, n \tag{5.3}$$

- **Pipelines**

Pipelines in the natural gas system are like superhighways for transporting natural gas from production areas to consumers. They are typically made of strong materials like steel and are buried underground to protect them. The gas flow is crucial in the system to ensure a steady supply of natural gas for various needs, from heating homes to powering industries. Key characteristics include their vast network spanning thousands of miles, their ability to transport large volumes of gas quickly and efficiently, and their reliability in delivering gas to homes, businesses, and industries. Pipelines play a crucial role in ensuring a steady supply of natural gas for various needs, from heating homes to powering industries.

As stated, the generation points and consumption ones are connected through pipelines which necessitates the use of gas flow by considering the pipelines characteristics. In this case, the most frequently used equation for this purpose is the Weymouth model (Wang et al., 2018) as presented in Eq. (5.4). In this equation, F_p is the flow through pipeline p and φ_p is the Weymouth constant. Moreover, Eq. (5.5) ensures the safe volume of the gas through pipeline p.

$$F_p^2 = \varphi_p\left(\rho_n^2 - \rho_m^2\right) \quad \forall m, n, p \tag{5.4}$$

$$-F_p^{\max} \le F_n \le F_n^{\max} \quad \forall p \tag{5.5}$$

- **Nodal gas balance**

The nodal gas balance is vital in the natural gas system as it ensures that the supply and demand of natural gas are balanced at various points within the network. Essentially, it is about matching the production, storage, and consumption of gas to maintain system stability and reliability. By accurately monitoring and managing gas flows at different nodes in the system, operators can prevent imbalances that could lead to disruptions or inefficiencies. This balance helps optimize resource utilization, minimize waste, and ensure a continuous and reliable supply of natural gas to meet the needs of consumers, industries, and other users. Accordingly, the nodal gas balance equation is represented as follows:

$$\sum_{p\in\text{Receive}} F_p - \sum_{p\in\text{Send}} F_p + \sum_{e} F_e^{\text{P2G}} + \sum_{w} F_w^{\text{Well}} + F_n^{\text{LS}} = F_n^L + \sum_{gf} F_{gf}^{\text{G2P}} \quad \forall n \tag{5.6}$$

where F_n^L and F_n^{LS} are the gas load and load shedding in natural gas system, respectively. F_e^{P2G} and F_{gf}^{G2P} represent the gas generation and consumption of P2G and G2P units, respectively.

5.4.2 *Power Systems Formulation*

Power systems can be represented as nodal systems, which include power generation units, transmission lines, and loads (for simplicity). Given that power transmission systems are well-known, the complete mathematical model for their operation is represented in Eqs. (5.7)–(5.11).

$$\sum_{l \in \text{Receive}} \text{PF}_l - \sum_{p \in \text{Send}} \text{PF}_l - \sum_e P_e^{\text{P2G}} + \sum_c P_c + P_b^{\text{LS}} = P_b^L - \sum_{gf} P_{gf}^{\text{G2P}} \quad \forall b \tag{5.7}$$

$$\text{PF}_l = B_l(\theta_b - \theta_d) \quad \forall (l, b, d) \tag{5.8}$$

$$-\overline{\text{PF}}_l \leq \text{PF}_l \leq \overline{\text{PF}_l} \quad \forall l \tag{5.9}$$

$$P_c^{\min} \leq P_c \leq P_c^{\max} \quad \forall c \tag{5.10}$$

$$0 \leq P_b^{\text{LS}} \leq P_b^L \quad \forall b \tag{5.11}$$

Equality (5.7) represents the power balance at each bus in the power system, where P_c is the generation from conventional generators, P_b^{LS} is the amount of load shedding, and P_b^L is the demand at bus *b*. In power transmission systems, the *X/R* ratio of the transmission lines is typically large, which allows for the use of DC power flow as shown in Eq. (5.8) (Li et al., 2022), where θ is the voltage angle at bus *b*. Finally, Eqs. (5.10) and (5.11) constrain the power generation of conventional generators and the amount of load shedding, respectively.

The optimization problem presented in Eq. (5.13) is an MINLP model which is a non-convex model and is intractable due to Eq. (5.4) (Li et al., 2022; Wang et al., 2018). Therefore, techniques to linearize or convexify the model are required and will be discussed in the subsequent section.

5.4.3 Coupling Devices Formulation

The coupling devices shown in Fig. 5.1 are P2G and G2P units. The P2G technology mainly employed in order to take full is ability in operating and extending the lifespan of renewable energy generation systems by eliminating wastage and enhancing operational efficiency over the long term (Shabanian-Poodeh et al., 2024a). This makes it possible to turn otherwise wasted power into gas in forms of hydrogen or methane which are easier to store as fuels rather than as electricity (Hu et al., 2024; Li et al., 2023). Conversely, G2P units enable the conversion of natural gas into power, for power generation purposes whenever necessary enabling easy energy sharing between both systems interconnecting them with perfect energy exchange in both directions (Shabanian-Poodeh et al., 2024a).

Equation (5.14) computes the energy conversion of the P2G units [the same applies to G2Ps in Eq. (5.16)], while Eq. (5.15) constrains the permissible operation of the P2Gs [likewise for G2Ps in Eq. (5.17)].

$$F_e^{\text{P2G}} = \frac{\mu_e^{\text{P2G}} P_e^{\text{P2G}}}{\text{GHV}} \tag{5.12}$$

$$F_{e,\min}^{\mathrm{P2G}} \leq F_{e}^{\mathrm{P2G}} \leq F_{e,\max}^{\mathrm{P2G}} \quad \forall e \tag{5.13}$$

$$F_{gf}^{\mathrm{G2P}} = \frac{\mu_{gf}^{\mathrm{G2P}} P_{gf}^{\mathrm{G2P}}}{\mathrm{GHV}} \tag{5.14}$$

$$F_{gf,\min}^{\mathrm{G2P}} \leq F_{gf}^{\mathrm{G2P}} \leq F_{gf,\max}^{\mathrm{G2P}} \quad \forall gf \tag{5.15}$$

where μ_{e}^{P2G} and μ_{gf}^{G2P} represent the conversion efficiencies of P2G and G2P units, respectively.

After all, utilizing the objective function for the optimal operation of the IPGS is as Eq. (5.16).

$$\mathrm{Min}\, Z = \sum_{c} P_c C_c^{\mathrm{Conv}} + \sum_{w} F_w^{\mathrm{Well}} C_w^{\mathrm{Well}} + C^{\mathrm{LS}} \left[\sum_{b} P_b^{\mathrm{LS}} + \mathrm{GHV} \sum_{n} F_n^{\mathrm{LS}} \right] \tag{5.16}$$

where $C_{[.]}^{[.]}$ represents the costs and GHV is the high calorific value of the natural gas. Therefore, the complete form of the operation problem of the IPGS can be formulated as follows:

$$\begin{aligned} &\textbf{Min } (16) \\ &\quad \text{s.t.} \\ &\quad \textbf{Gas} \rightarrow (1)-(6) \\ &\quad \textbf{Power} \rightarrow (7)-(11) \\ &\quad \textbf{Couplings} \rightarrow (12)-(15) \end{aligned} \tag{5.17}$$

5.5 Model Convexification Techniques

In general, the presence of non-convexities and nonlinearities adds complexity to the optimization process as it results in intricate and irregular solution spaces with numerous local optima (Burer & Letchford, 2012). Overcoming these challenges requires the utilization of advanced algorithms and significant computational power to find the best solutions, or the use of convexification methods to simplify the computational workload. In the case of IPGS operation, the gas flow equation, also known as the Weymouth equation, is frequently the main cause of non-convexity. To tackle this issue, two popular approaches are piecewise linearization (PWL) and second-order cone relaxation (SOCR) techniques. In the following sections, we will provide a detailed explanation of each of these methods.

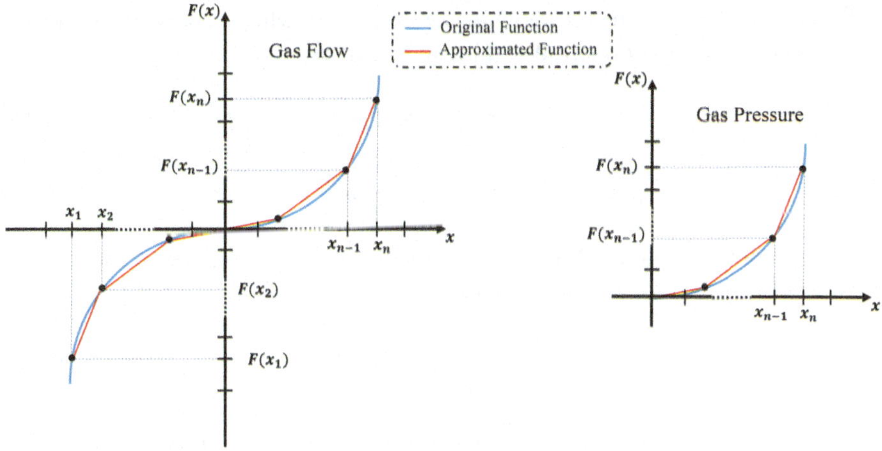

Fig. 5.4 Gas flow and pressure linearization using PWL

5.5.1 PWL Method

The PWL is a technique used to approximate a nonlinear function by breaking it down into linear segments within specific intervals (Lin et al., 2013), as shown in Fig. 5.4. This simplifies the problem and enables the use of mixed-integer linear programming (MILP). In gas flow, gas pressure is always positive, while gas flow in pipelines may be negative. To obtain a feasible and optimal solution, only one segment from the segmented linear curve is required, which is achieved by introducing a binary variable for each segment and a constraint that enforces the selection of just one segment. The PWL method has been used in References (Javadi et al., 2022; Sayed et al., 2019; Shabanian-Poodeh et al., 2025; Wang et al., 2023).

5.5.2 SOCR Technique

The second-order cone relaxation (SOCR) method is utilized in mathematical modeling and convex optimization to approximate and tackle optimization problems featuring non-convex quadratic constraints and objective functions. This technique serves as a potent tool for converting non-convex optimization problems into convex counterparts, enabling effective solutions using existing optimization algorithms (Wang et al., 2018). In an optimization problem, a second-order conic constraint is a specific type of constraint that encompasses a quadratic inequality. For instance, assuming the nonlinear and non-convex relationship is defined as follows:

$$a_{ij}^2 + b_{ij}^2 = a_{ii}.b_{jj} \tag{5.18}$$

This relationship can be reformulated as Eq. (5.19), which represents its convex hull (Liu et al., 2021). Furthermore, this can be expressed as Eq. (5.20).

$$a_{ij}^2 + b_{ij}^2 \leq a_{ii}.b_{jj} \tag{5.19}$$

$$\|2a_{ij}2b_{ij} - a_{ii}\| \leq b_{jj} \tag{5.20}$$

Here, $\|.\|$ represents the Euclidean norm (length) of the vector. Hence, Eq. (5.4) is transformed into a convex form as follows:

$$\|\frac{F_p}{\varphi_p.\rho_n}\| \leq \rho_m \quad \forall m, n, p \tag{5.21}$$

Upon applying this technique, the planning model is transformed into a quadratically constrained programming (SOCP) model. This convexification method has been applied in various references (Li et al., 2023; Lv et al., 2022). It is important to note a significant aspect of this method: there is potential for the resulting solution to be somewhat imprecise, without a guarantee of achieving the globally optimal answer (Li et al., 2020). To mitigate this and ensure the accuracy of the relaxation during the solution process, a linear penalty term should be integrated into the objective function (Li et al., 2020). The linear penalty that is added to the objective is the difference of the connected nodes (right-hand-side of Eq. (5.4) as $\sum_{n,m} \xi(\rho_n - \rho_m)$, where ξ is positive scalar. Additionally, relaxation deviation indices are introduced to assess the solution's accuracy, ideally approaching zero (Li et al., 2020).

5.6 Numerical Results

This section presents the results of a four-bus/four-node example, with the topology illustrated in Fig. 5.5. Bus bars, gas nodes, power and gas loads, natural gas wells, and generators are labeled consistently with the model formulation. The G2P and P2G unit 1 connect two systems. All relevant data are available in an online supplement (https://doi.org/10.6084/m9.figshare.6025340.v1). We analyze system operations in a normal (baseline) scenario and two additional scenarios where natural gas demands are surged by 10 and 20% compared to the normal load. The QCP model was coded in GAMS using Gurobi solver to illustrate the effect of presented model for optimal operation of IPGS. The QCP model implemented in the code incorporates the objective function (5.16), along with constraints defined by Eqs. (5.1)–(5.3) and (5.5)–(5.15). Additionally, Eq. (5.4) is replaced with Eq. (5.21) to transform the NLP model to the QCP one.

The numerical studies are conducted under three different scenarios:

- Baseline of natural gas demand,
- 10% increase in natural gas demand, and

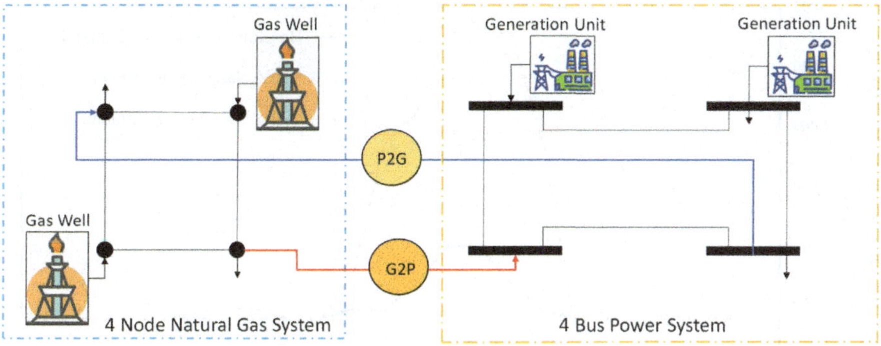

Fig. 5.5 System topology of the example

- 20% increase in natural gas demand.

The primary reason for selecting these scenarios is to effectively demonstrate the importance of coordination between interconnected energy systems and how changes in one system can propagate and impact the others.

5.6.1 Coordination of IPGS

Figure 5.6a, b depicts the interdependencies between natural gas and electric LMPs, highlighting the interconnected operating costs of both systems. Figure 5.6a displays hourly load-weighted natural gas LMPs derived from the SOCP. All three scenarios show high LMPs during hours 9–12, which are more pronounced in scenarios with increased natural gas demands. The elevated LMPs in these high-load-demand scenarios are attributed to congestion in the natural gas system, which leads to curtailment of natural gas demand. This load shedding increases the operating costs of the G2P units during these periods, resulting in the higher electric LMPs observed in Fig. 5.6b.

The high natural gas LMPs during hours 9–12 further impact the system by reducing the utilization of natural gas-fired generation. In the baseline scenario, about 15 GWh of G2P generation is produced throughout the modeled day. This amount decreases to 13 GWh and 9 GWh in scenarios where natural gas loads are 10% and 20% above the baseline, respectively. Consequently, congestion in the natural gas system limits the use of G2P units and increases electric LMPs. Additionally, the objective function values and solution times for two solvers, CPLEX and Gurobi, are presented in Table 5.1. As shown, the optimal objective function values for the three scenarios are nearly identical, but Gurobi demonstrated slightly faster performance compared to CPLEX.

The results indicate that the coordination between natural gas and electric power systems is crucial because of their interdependency. Natural gas is a key fuel for

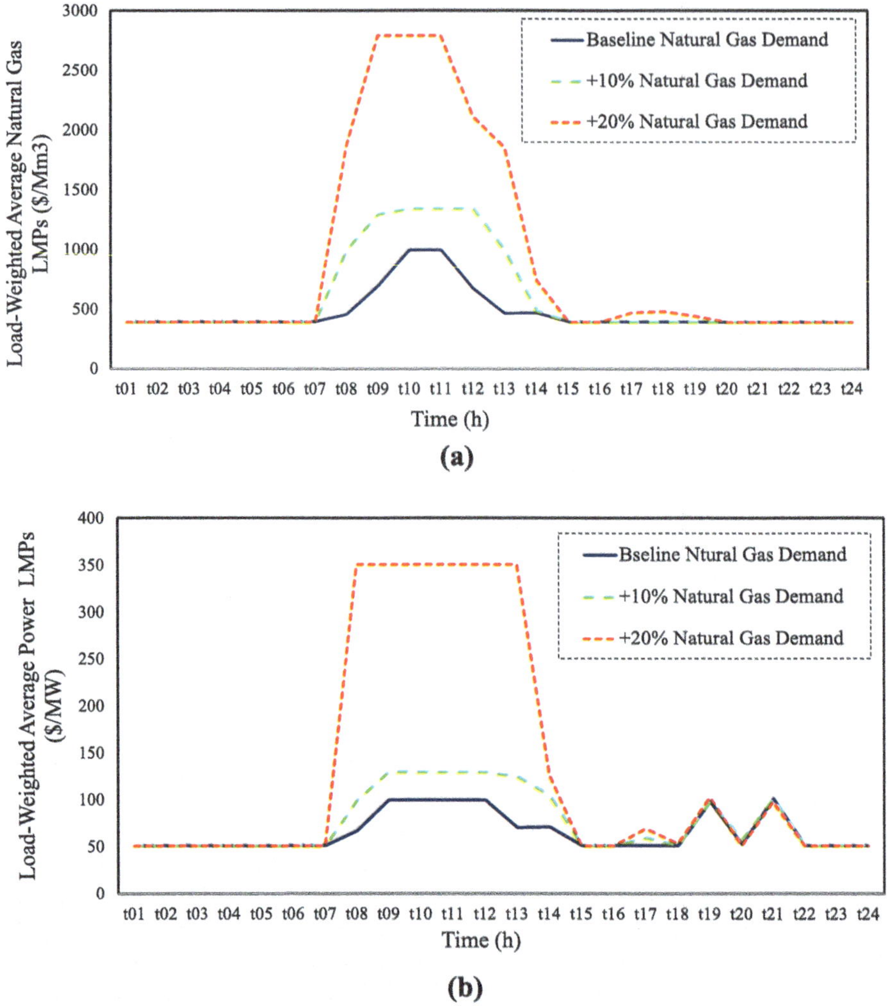

Fig. 5.6 Load-weighted average LMPs of **a** natural gas and **b** power system

Table 5.1 Objective function value and solution time for three scenarios

Solver	Objective function value (M$)			Solution time (s)
	Baseline	10% higher	20% higher	
CPLEX	3.832	4.215	4.702	1.3
Gurobi	3.832	4.22	4.697	1.26

electricity generation, and any congestion or disruption in the natural gas system directly affects the availability and cost of electricity. Without proper coordination, natural gas curtailment can cause inefficiencies in power generation, leading to higher electricity prices, reduced reliability, and a potential mismatch between supply and demand.

Effective IPGS coordination helps in:

(1) **Mitigating Congestion and Curtailment**: Coordinated planning and operation can prevent bottlenecks in the natural gas system, reducing the need for demand curtailment and maintaining stable operating costs.
(2) **Stabilizing Prices**: Improved coordination ensures that fluctuations in one system do not disproportionately affect the other, stabilizing both natural gas and electric LMPs.
(3) **Enhancing Reliability and Resilience**: Close coordination enhances the ability to manage high-demand scenarios, thereby improving the overall resilience of both gas and electric systems. During disruptions caused by natural events in one system, the other can provide support to maintain maximum load through coupling devices, helping to effectively navigate the catastrophic situation.

5.6.2 Computational Performance Assessment

As discussed in Sect. 4.2, the SOCR relaxation must be evaluated to ensure both feasibility and global optimality of the results. Since the linear penalty function, outlined in Sect. 4.2, has been incorporated into the objective function to minimize deviations, a computational assessment of the relaxation method is necessary. To quantify the maximum errors resulting from the relaxation in Eq. (5.21) over the time period, Index (5.22) is used as follows (Li et al., 2020, 2022; Shabanian-Poodeh et al., 2024b):

$$Error = \underset{p}{\mathrm{Max}}\left\{\left|\rho_m^* \quad \left(\left[\frac{F_p^*}{\varphi_p}\right]^2 - \left[\rho_n^*\right]^2\right)\right|\right\} \tag{5.22}$$

where in this equation, * means the obtained value of the variable. According to Fig. 5.7, the maximum deviation occurred is 0.094×10^{-3} at time interval 3. The error is negligible, making it reasonable to consider the obtained results as globally optimal.

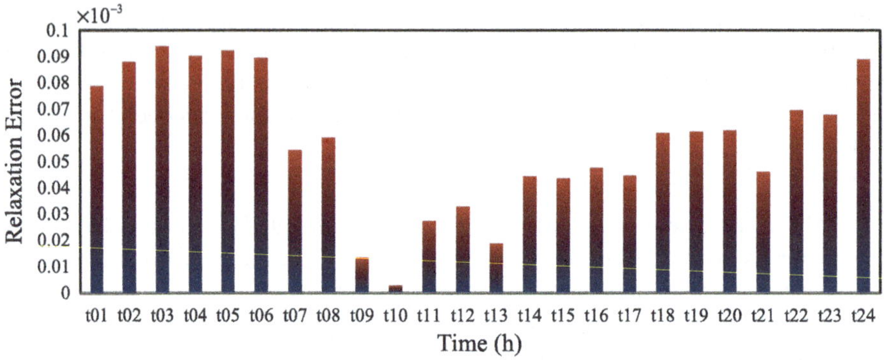

Fig. 5.7 Maximum deviation of gas flow equation relaxation

5.7 Challenges and Limitations in IPGS Implementation

Still, there are obvious challenges that come with the real-world implementation of IPGS. System complexity is one of the main issues because of the underlying complexity in real-time integration of power and gas networks. For example, gas may be used for both electricity production and residential heating during peak demand times, which often leads to conflicting objectives. Data quality and availability are another issue, when there are inconsistencies in standard measurement units or reporting periods of one power system and other gas systems, which may cause poor decisions. Additionally, regulatory frameworks may not adequately address the integrated nature of IPGS, creating barriers to efficient operation. Different pricing mechanisms for electricity and gas markets, for example, can hinder optimal resource allocation across the integrated system.

Current convexification methods used in IPGS optimization also have notable limitations. These techniques often trade off accuracy for computational speed, which can be problematic in real-time operations. Simplified convex models of gas flow equations, for instance, may not capture pressure dynamics accurately, leading to suboptimal or infeasible solutions in practice. Moreover, convexified models may perform poorly under extreme operating conditions or during contingencies, such as the sudden loss of a major gas pipeline. Scalability is another concern, as these methods may struggle to maintain acceptable solution quality and computational time as IPGS grows in size and complexity. The integration of a large number of distributed energy resources and gas storage facilities can exponentially increase problem complexity, challenging the effectiveness of existing convex relaxations. Lastly, convexification techniques often require careful tuning for specific system configurations, limiting their adaptability to dynamic changes in network topology and making it difficult to incorporate uncertainties from renewable generation, demand fluctuations, and market prices.

5.8 Conclusion

In conclusion, integrated power-gas systems (IPGSs) offer a transformative approach to energy infrastructure, integrating power, and gas energy systems to establish resilient and sustainable energy ecosystems. The coordination of power-to-gas (P2G) and gas-to-power (G2P) processes lies at the core of IPGS, facilitating the conversion of excess renewable energy into storable gases and their subsequent reconversion into electricity as needed. This symbiotic relationship is managed by a central coordinator to ensure efficient resource utilization and dynamic balancing of supply and demand. The optimal operation of IPGS relies on mathematical optimization models that consider system limitations and characteristics. However, non-convex constraints often pose computational challenges, necessitating convexification methods such as piecewise linearization (PWL) and second-order cone relaxation (SOCR). These techniques make the models more tractable, enabling their application to real-world problems, particularly in large-scale planning scenarios.

Resilience is a crucial aspect of IPGS, ensuring continuous and reliable energy supply despite disruptions. Enhancing resilience involves strategic planning, coupling devices like P2G and G2P, and defining clear resilience criteria for assessment. Long-term measures such as expansion planning and redundancy incorporation strengthen IPGS against extreme events, while operational criteria focus on maintaining optimal performance during disturbances. Another standpoint of view for the sustainable supply chain in the context of the IPGS is the market which was discussed in this chapter. Convexification techniques like PWL and SOCR address non-convexities in optimization models, making them more manageable. These methods play a vital role in optimizing the operation of IPGS, ensuring efficiency, reliability, and sustainability.

We investigated the interdependencies between prices in the two systems and found that congestion in one system can impact prices in the other, with this effect potentially being bidirectional. For example, as shown in Sect. 5.5, high natural gas load can force the curtailment of natural gas loads, significantly increasing natural gas LMPs. This issue makes G2Ps generation more costly, decreasing their use while simultaneously increasing electric LMPs. Additionally, the computational performance of the QCP model was visualized and analyzed to present a comprehensive view for the readers.

In summary, IPGS offers a promising solution to the challenges of energy infrastructure, providing resilience, sustainability, and efficient resource utilization. By integrating power and gas systems and employing advanced optimization techniques, IPGS paves the way for a cleaner, more reliable energy future.

The IPGS presents a complex and dynamic field with numerous opportunities for future research and advancement. Building upon the findings of this study, we identify several promising areas for further investigation:

1. **Advanced Convexification Techniques**: Exploring hybrid approaches that combine convex optimization with machine learning to improve accuracy and computational efficiency.

2. **AI and Machine Learning Integration**: The application of artificial intelligence and machine learning algorithms shows great promise in optimizing IPGS operations. Specifically, predictive analytics could be employed to forecast energy demand and supply fluctuations, while reinforcement learning techniques could be used to develop adaptive control strategies for more robust system management.
3. **Multi-vector Energy Storage**: Investigating the optimal integration and operation of various energy storage technologies across both power and gas systems could significantly enhance system flexibility and resilience. This includes exploring synergies between batteries, hydrogen storage, and compressed air energy storage.
4. **Cyber-physical Security**: As IPGS becomes increasingly digitalized, ensuring their security against cyber-attacks is paramount. Future research should focus on developing advanced cybersecurity measures and resilient control strategies to protect these critical infrastructures.

References

Aldarajee, A. H. M., Hosseinian, S. H., Vahidi, B. (2020). A secure tri-level planner-disaster-risk-averse replanner model for enhancing the resilience of energy systems. *Energy, 204*. https://doi.org/10.1016/j.energy.2020.117916

Armioun, M., Nazar, M. S., Shafie-khah, M., & Siano, P. (2023). Optimal scheduling of CCHP-based resilient energy distribution system considering active microgrids' multi-carrier energy transactions. *Applied Energy, 350*, 121719. https://doi.org/10.1016/j.apenergy.2023.121719

Burer, S., & Letchford, A. N. (2012). Non-convex mixed-integer nonlinear programming: A survey. *Surveys in Operations Research and Management Science, 17*(2), 97–106. https://doi.org/10.1016/j.sorms.2012.08.001

Chen, M., Lu, H., Chang, X., & Liao, H. (2023). An optimization on an integrated energy system of combined heat and power, carbon capture system and power to gas by considering flexible load. *Energy, 273*, 127203. https://doi.org/10.1016/j.energy.2023.127203

Daneshvar, M., Mohammadi-Ivatloo, B., & Zare, K. (2023). A fair risk-averse stochastic transactive energy model for 100% renewable multi-microgrids in the modern power and gas incorporated network. *IEEE Transactions on Smart Grid, 14*(3), 1933–1945. https://doi.org/10.1109/TSG.2022.3218255

Dimitriadis, C. N., Tsimopoulos, E. G., & Georgiadis, M. C. (2023a). Optimal bidding strategy of a gas-fired power plant in interdependent low-carbon electricity and natural gas markets. *Energy, 277*, 127710. https://doi.org/10.1016/j.energy.2023.127710

Dimitriadis, C. N., Tsimopoulos, E. G., & Georgiadis, M. C. (2023b). Optimization-based economic analysis of energy storage technologies in a coupled electricity and natural gas market. *J Energy Storage, 58*, 106332. https://doi.org/10.1016/j.est.2022.106332

Faramawy, S., Zaki, T., & Sakr, A. A.-E. (2016). Natural gas origin, composition, and processing: A review. *Journal of Natural Gas Science and Engineering, 34*, 34–54. https://doi.org/10.1016/j.jngse.2016.06.030

Gasser, P., et al. (2021). A review on resilience assessment of energy systems. *Sustain Resilient Infrastruct, 6*(5), 273–299. https://doi.org/10.1080/23789689.2019.1610600

Gharai, M. A., Hooshmand, R.-A., & Nosratabadi, S. M. (2023). Eco-environmental uncertain management of various energy hubs within a microgrid concentrated on demand response, intelligent parking lots, and power-to-gas unit effects. *International Journal of Energy Research, 2023*, 1–30. https://doi.org/10.1155/2023/3173099

Hasanzad, F., & Rastegar, H. (2022). Application of optimal hardening for improving resilience of integrated power and natural gas system in case of earthquake. *Reliability Engineering and System Safety, 223*, 108476. https://doi.org/10.1016/j.ress.2022.108476

He, C., Dai, C., Wu, L., & Liu, T. (2018). Robust network hardening strategy for enhancing resilience of integrated electricity and natural gas distribution systems against natural disasters. *IEEE Transactions on Power Systems, 33*(5), 5787–5798. https://doi.org/10.1109/TPWRS.2018.2820383

Hemmati, R., Mehrjerdi, H., & Nosratabadi, S. M. (2021). Resilience-oriented adaptable microgrid formation in integrated electricity-gas system with deployment of multiple energy hubs. *Sustainable Cities and Society, 71*. https://doi.org/10.1016/j.scs.2021.102946

https://doi.org/10.6084/m9.figshare.6025340.v1

Hu, J., Zou, Y., & Zhao, Y. (2024). Robust operation of hydrogen-fueled power-to-gas system within feasible operating zone considering carbon-dioxide recycling process. *International Journal of Hydrogen Energy, 58*, 1429–1442. https://doi.org/10.1016/j.ijhydene.2024.01.337

Javadi, E. A., Joorabian, M., & Barati, H. (2022). A sustainable framework for resilience enhancement of integrated energy systems in the presence of energy storage systems and fast-acting flexible loads. *Journal of Energy Storage, 49*, 104099. https://doi.org/10.1016/j.est.2022.104099

Jia, K., Liu, C., Li, S., & Jiang, D. (2023). Modeling and optimization of a hybrid renewable energy system integrated with gas turbine and energy storage. *Energy Conversion and Management, 279*, 116763. https://doi.org/10.1016/j.enconman.2023.116763

Jufri, F. H., Widiputra, V., & Jung, J. (2019). State-of-the-art review on power grid resilience to extreme weather events: Definitions, frameworks, quantitative assessment methodologies, and enhancement strategies. *Applied Energy, 239*, 1049–1065. https://doi.org/10.1016/j.apenergy.2019.02.017

Kidnay, A. J., Parrish, W. R., & McCartney, D. G. (2019). Fundamentals of natural gas processing. *CRC Press*. https://doi.org/10.1201/9780429464942

Kocuk, B., Dey, S. S., & Sun, X. A. (2016). Strong SOCP relaxations for the optimal power flow problem. *Operations Research, 64*(6), 1177–1196. https://doi.org/10.1287/opre.2016.1489

Kostowski, W. J., Kalina, J., Bargiel, P., & Szufleński, P. (2015). Energy and exergy recovery in a natural gas compressor station—A technical and economic analysis. *Energy Conversion and Management, 104*, 17–31. https://doi.org/10.1016/j.enconman.2015.07.002

Li, G., Yan, K., Zhang, R., Jiang, T., Li, X., & Chen, H. (2022). Resilience-oriented distributed load restoration method for integrated power distribution and natural gas systems. *IEEE Transactions on Sustainable Energy, 13*(1), 341–352. https://doi.org/10.1109/TSTE.2021.3110975

Li, P. (2022). Techno-environmental assessment of gas-based power plant and renewable unit in the presence of power to gas system and gas storages for emission reduction using information gap decision theory. *Journal of Energy Storage, 51*, 104382. https://doi.org/10.1016/j.est.2022.104382

Li, S., Ding, T., Mu, C., Huang, C., & Shahidehpour, M. (2022). A machine learning-based reliability evaluation model for integrated power-gas systems. *IEEE Transactions on Power Systems, 37*(4), 2527–2537. https://doi.org/10.1109/TPWRS.2021.3125531

Li, T., Han, X., Wu, W., & Sun, H. (2023). Robust expansion planning and hardening strategy of meshed multi-energy distribution networks for resilience enhancement. *Applied Energy, 341*. https://doi.org/10.1016/j.apenergy.2023.121066

Li, X., et al. (2023). Operation optimization for integrated energy system based on hybrid CSP-CHP considering power-to-gas technology and carbon capture system. *Journal of Cleaner Production, 391*, 136119. https://doi.org/10.1016/j.jclepro.2023.136119

Li, Y., Li, Z., Wen, F., & Shahidehpour, M. (2020). Minimax-regret robust co-optimization for enhancing the resilience of integrated power distribution and natural gas systems. *IEEE Transactions on Sustainable Energy, 11*(1), 61–71. https://doi.org/10.1109/TSTE.2018.2883718

Lin, M.-H., Carlsson, J. G., Ge, D., Shi, J., & Tsai, J.-F. (2013). A review of piecewise linearization methods. *Mathematical Problems in Engineering, 2013*, 1–8. https://doi.org/10.1155/2013/101376

Liu, R.-P., Sun, W., Yin, W., Zhou, D., & Hou, Y. (2021). Extended convex hull-based distributed optimal energy flow of integrated electricity-gas systems. *Applied Energy, 287*, 116551. https://doi.org/10.1016/j.apenergy.2021.116551

Lv, C., Liang, R., Jin, W., Chai, Y., Yang, T.: Multi-stage resilience scheduling of electricity-gas integrated energy system with multi-level decentralized reserve. *Applied Energy, 317*. https://doi.org/10.1016/j.apenergy.2022.119165

Moradi, A., Salehi, J., & Shafie-khah, M. (2024). An interactive framework for strategic participation of a price-maker energy hub in the local gas and power markets based on the MPEC method. *Energy, 307*, 132608. https://doi.org/10.1016/j.energy.2024.132608

Nasiri, N., et al. (2023). A robust bi-level optimization framework for participation of multi-energy service providers in integrated power and natural gas markets. *Applied Energy, 340*, 121047. https://doi.org/10.1016/j.apenergy.2023.121047

Sayed, A. R., Wang, C., & Bi, T. (2019). Resilient operational strategies for power systems considering the interactions with natural gas systems. *Applied Energy, 241*, 548–566. https://doi.org/10.1016/j.apenergy.2019.03.053

Shabanian-Poodeh, M., Hooshmand, R.-A., & Kabiri-Renani, Y. (2024a). Probabilistic co-expansion planning for natural gas and electricity energy systems with wind curtailment mitigation considering uncertainties. *IET Generation Transmission & Distribution*. https://doi.org/10.1049/gtd2.13268

Shabanian-Poodeh, M., Hooshmand, R.-A., & Shafie-khah, M. (2024b). Risk-constrained strategic resilience enhancement of power distribution and natural gas systems through G2P and P2G planning and distributed load restoration against flood. *Sustainable Cities and Society*, 105618. https://doi.org/10.1016/j.scs.2024.105618

Shabanian-Poodeh, M., Hooshmand, R.-A., & Shafie-khah, M. (2025). Reliability-constrained configuration optimization for integrated power and natural gas energy systems: A stochastic approach. *Reliability Engineering and System Safety, 254*, 110600. https://doi.org/10.1016/j.ress.2024.110600

Wang, C., Wei, W., Wang, J., Bai, L., Liang, Y., & Bi, T. (2018). Convex optimization based distributed optimal gas-power flow calculation. *IEEE Transactions on Sustainable Energy, 9*(3), 1145–1156. https://doi.org/10.1109/TSTE.2017.2771954

Wang, H., Hou, K., Zhao, J., Yu, X., Jia, H., & Mu, Y. (2022). Planning-oriented resilience assessment and enhancement of integrated electricity-gas system considering multi-type natural disasters. *Applied Energy, 315*. https://doi.org/10.1016/j.apenergy.2022.118824

Wang, Y., Yang, Y., & Xu, Q. (2023). Integrated planning of natural gas and electricity distribution systems for enhancing resilience. *International Journal of Electrical Power & Energy Systems, 151*, 109103. https://doi.org/10.1016/j.ijepes.2023.109103

Wang, Z., et al. (2022). Multi-stage stochastic programming for resilient integrated electricity and natural gas distribution systems against typhoon natural disaster attacks. *Renewable and Sustainable Energy Reviews, 159*, 111784. https://doi.org/10.1016/j.rser.2021.111784

Xie, H., Sun, X., Chen, C., Bie, Z., & Catalao, J. P. S. (2022). Resilience metrics for integrated power and natural gas systems. *IEEE Transactions on Smart Grid, 13*(3), 2483–2486. https://doi.org/10.1109/TSG.2022.3145635

Xu, Y., et al. (2023). A review of resilience enhancement strategies in renewable power system under HILP events. *Energy Reports, 9*, 200–209. https://doi.org/10.1016/j.egyr.2022.12.128

Zandrazavi, S. F., Tabares, A., Franco, J. F., Shafie-khah, M., Soares, J., & Vale, Z. (2023). A two-stage stochastic model for coordinated operation of natural gas and microgrid networks. In *2023 IEEE Belgrade PowerTech. IEEE 2023* (pp. 01–07). https://doi.org/10.1109/PowerTech55446.2023.10202794

Zhang, Y., & Ding, Z. (2023). Full-time scale resilience enhancement strategies for an integrated electricity-gas-transportation coupled system. *Applied Mathematical Modelling, 117*, 63–86. https://doi.org/10.1016/j.apm.2022.12.020

Chapter 6
Challenges in Renewable Energy Supply Chains: Strategies for a Sustainable Energy Future Authors

Saleem ur Rahman and Shah Rukh Shakeel

Abstract Mitigating climate change is one of the most pressing issues today. A key step in addressing climate change is reducing reliance on fossil fuels, which requires a global shift toward renewable energy. However, this transition faces numerous supply chain challenges, including unpredictable resource availability, financial limitations, regional political instability, technological gaps, social and environmental impacts, and inconsistent regulations and policies. These challenges lead to production delays, power interruptions, decreased energy efficiency, and disruption of the smooth flow of materials, slowing the global adoption of renewable energy technologies. This chapter addresses these critical challenges in the renewable energy sector. The findings indicate that geographical, technological, geopolitical, infrastructural, social and environmental, financial, regulatory, and governmental challenges significantly affect the entire supply chain, from extraction to the production of renewable energy technologies between technology-rich and resource-rich countries. The chapter also suggests future trends for overcoming supply chain disruptions and fostering greater cooperation, innovation, and sustainability in transitioning to a renewable energy-based future.

Keywords Renewable energy · Supply chain · Challenges · Energy transition · Sustainability

---

S. Rahman (✉)
Rabat Business School, The International University of Rabat (Université Internationale de Rabat), Parc Technopolis, Shore Rocade, 11103 Rabat, Morocco
e-mail: rahman.saleem@uir.ac.ma

S. R. Shakeel
School of Marketing and Communication, and Innovation and Entrepreneurship InnoLab, University of Vaasa, Wolffintie 34, Vaasa, Finland
e-mail: shah.rukh.shakeel@uwasa.fi

I. Rahimi et al. (eds.), *Sustainable Supply Chain of Renewable Energy Networks*, Energy Systems in Electrical Engineering, https://doi.org/10.1007/978-981-96-4940-2_6

6.1 Introduction

Climate change affects every region worldwide and will cost more than preventing its negative impacts (International Renewable Energy Agency (IRENA), 2021). In recent years, human activities have set new records for greenhouse gas emissions (GHG), rising sea levels, increased waste, ocean acidification, and deforestation (United Nations (UN), 2024a). The Intergovernmental Panel on Climate Change (IPCC) has called for immediate and strict actions to address these accelerating climate change issues (IPCC, 2018). Conventional hydrocarbons, including coal, oil, and gas, are the largest contributor to climate change, accounting for over 90% of carbon emissions and over 75% of global greenhouse gas emissions (UN, 2024b). Therefore, to mitigate the worst effects of climate change, carbon emissions must be cut by nearly half by 2030 from the current levels and should reach net zero levels by 2050. Transitioning to clean energy systems based on renewable energy sources such as solar, wind, hydro, and other renewables offers a promising path to a more sustainable future (Yu et al., 2024; Paris Agreement, 2024).

Renewable energy sources (RETs) are sustainable, clean, affordable, reliable, and accessible (United Nations Environment Programme (UNEP), 2024; Shakeel et al., 2017). Currently, renewable energy sources play a significant role in addressing some countries' primary energy needs. However, despite these developments, their overall contribution to global energy remains limited. Nevertheless, RETs are emerging as a promising alternative to conventional hydrocarbons, which are associated with numerous issues. Excessive reliance on finite hydrocarbons, emissions from generation processes, the need for massive infrastructural investments, concerns over energy access and poverty, and energy security are some of the primary challenges plaguing conventional energy systems. These challenges have prompted countries worldwide to accelerate the development of renewable energy technologies and increase their share in primary energy supplies.

Technological advancements and improved efficiency at lower costs support the expansion of renewable energy generation, while supportive policy instruments have also been introduced to stimulate growth. As a result, the contribution of renewables to global energy supplies is projected to increase significantly in the coming years. According to IRENA (2024), 29.1% of the existing power generation facilities installed worldwide in 2022 were based on renewable sources, totaling 8440 TWh, a 7.2% increase from 2021. Altogether, over $2 trillion has been invested in the cleantech sector (IEA, 2024), with more than $619 billion directed explicitly toward renewable energy initiatives globally (Statista, 2024). Furthermore, more than 130 countries have developed various policy supports or regulations to promote the growth of renewable energy technologies (REN21, 2024; IEA, 2024).

However, to reach the stated target and increase the share of renewables, the production of renewable energy technologies must be continuously scaled up. A significant technical challenge associated with the development of RETs is their heavy reliance on a wide range of materials and minerals, such as lithium, cobalt, nickel, and other rare earth elements (IEA, 2022), which are essential for the development of technologies and components like solar panels, wind turbines, and storage systems. The global distribution of these resources is highly uneven, with many resource-rich countries, in terms of mineral deposits, being geographically distant from the industrial centers where RETs are manufactured. For instance, cobalt is predominantly extracted from the Democratic Republic of Congo, while major lithium reserves are found in Chile and Australia.

This geographical disparity creates significant challenges for the supply chain, as manufacturing regions such as North America, Europe, and East Asia depend on long-distance transportation of materials. This reliance heightens the risk associated with the supply chain, including geopolitical tensions, trade wars or restrictions, complex and costly logistics, transparency issues, and safety and security concerns. These factors can lead to supply chain disruptions and price volatility, ultimately affecting renewable energy markets' development, growth, and stability. With the increasing utilization of RETs, the demand for essential materials will increase rapidly, making achieving sustainable development goals more complex and highly dependent upon an efficient supply chain. Interruptions in the supply of renewable energy materials can lead to significant delays in development (World Future Council, 2024).

This growing demand for renewable energy sources underscores the evolving and interdependent relationship between resource-rich and technology-advanced countries, which is essential for ensuring a smooth transition to clean energy (Feng et al., 2024). Building on this background, this chapter focuses on the complex issues surrounding the extraction, supply, and delivery of renewable energy materials. It addresses the geographical, logistical, geopolitical, financial, social, environmental, and policy challenges that impact both resource-rich and technologically advanced countries. Additionally, it examines the crucial role and effects of market dynamics, trade regulations, and technological breakthroughs in shaping the distribution and supply of renewable energy materials. Lastly, the chapter discusses strategies and suggestions to overcome supply chain barriers, promoting resilience and enabling a more effective transition within the renewable energy industry.

6.2 Challenges in the Renewable Energy Supply Chain

The development of renewable energy technologies encounters numerous challenges within the supply chain that hinder the smooth transition to a sustainable energy future. These challenges range from the sourcing and availability of critical materials to broader issues such as technological discrepancies, logistical and infrastructural constraints, geopolitical tensions, and governmental challenges. Social and environmental concerns, policy and regulatory challenges, and financial considerations further complicate supply chain dynamics. Each of these factors plays a critical role

Fig. 6.1 Challenges in renewable energy supply chain networks (Authors' work)

in influencing the effectiveness and sustainability of the supply chain, both independently and in combination with others. The following section explores these challenges, their impact on renewable energy supply chains (see Fig. 6.1), and measures to address them (see Fig. 6.2).

6.2.1 Geographical Disparities

Geographical disparities in the renewable energy supply chain refer to the uneven distribution of energy resources and the geographical distances between resource-rich and resource-deprived regions. Countries across the globe possess varying degrees of energy resources, and their transition to a clean energy future is heavily influenced by the uneven distribution of these resources and the technologies required to harness them (REN21, 2021). This dynamic creates a complex interplay between resource-rich and technologically advanced countries (Feng et al., 2023). Resource-abundant countries often export renewable energy materials, while resource-scarce countries depend on imports to meet their needs. This interdependence between resource-rich nations and technologically advanced countries, with knowledge, skills, capacities, and expertise to develop RETs, significantly shapes the global energy landscape, driving trade patterns and influencing policies worldwide.

Fig. 6.2 Addressing supply chain network challenges (Authors' work)

A significant challenge in this global energy landscape is the geographical distance between resource-rich and technology-intensive regions (Bridge et al., 2013). The largest importers of renewable materials include the European Union (EU), China, the United States, Japan, and South Korea (WTO, 2024). On the other hand, countries like Australia (lithium), Chile (copper and lithium), the Democratic Republic of Congo (cobalt), Indonesia (nickel), and South Africa (platinum and iridium resources) dominate the supply side (IRENA, 2021). This geographical disparity results in significant logistical challenges, as these resources must travel long distances, adding financial and operational complexity to the supply chain.

In some cases, the proximity of resource-rich and technology-rich countries creates a natural pull effect for renewable energy materials. For instance, countries like Germany, Spain, Italy, and France, known for their excellence in RET development (e.g., solar panels and wind turbines), actively invest in and collaborate with resource-rich nations to secure necessary materials. By developing partnerships and collaborations, they can leverage resources from nearby regions, such as Africa, to meet their technology development needs (Samadi et al., 2023). Countries that are close to each other have a natural comparative advantage in the energy supply chain, as shorter distances reduce logistical costs and ensure quicker access to resources.

Several strategies can be employed to mitigate the challenges posed by geographical disparities. Strengthening international relations through bilateral agreements can foster cooperation between resource-rich and technologically advanced countries, facilitating the sharing and management of resources and technologies for mutual benefit. Additionally, enhancing transparency in trade policies can minimize barriers to fair trade in energy resources. Supporting the development of regional processing hubs near extraction sites to refine material before long-distance transportation can help manage the volume and costs, offsetting the challenges posed by geographical distances. Moreover, adopting innovative supply chain management approaches, including digitalization and novel technological solutions, can further address challenges raised by geographical distances. While geographical disparities pose significant challenges to the renewable energy supply chain, they also present opportunities for collaboration and infrastructure development. By implementing these strategies, countries can address the challenges associated with geographical distances and work toward a more efficient and resilient supply chain.

6.2.2 Technological Discrepancies

The efficient, sustainable, and steady extraction, supply, distribution, and availability of raw materials are crucial for manufacturing renewable energy technologies. The integration of digital technology is essential in optimizing these processes. Technology is instrumental in the transition to cleaner energy by creating new opportunities in the energy industry. For example, the integration and use of digital technologies not only reduce operational costs and the carbon footprint of supply chain networks but also improve efficiencies and enhance safety aspects.

Advanced technologies also enable enhanced visibility by offering real-time tracking and monitoring of goods, maintaining inventory levels, and optimizing transportation routes. Advanced algorithms and machine learning models can optimize demand forecasts, contributing to efficient inventory management (Li et al., 2023). Similarly, they increase agility and responsiveness, promoting quicker decision-making and faster responses in the event of disruptions or changes in market dynamics. Additionally, advanced technologies enable improved interaction with customers through digital interfaces and personalized solutions. They also ensure that quality standards are maintained using IoT sensors and data analytics to monitor quality throughout the supply chain. The use of digital platforms enhances collaboration and communication between stakeholders.

Digital technologies such as artificial intelligence (AI), machine learning (ML), blockchain, the Internet of Things (IoT), and cloud computing can be promising for optimizing supply chains, improving renewable energy product lifecycle management, and enhancing efficiency, including providing economic, social, and environmental benefits of using renewable energy sources (Almutairi et al., 2023; Sánchez-García et al., 2024). The integration of technologies can also help avoid delays, shortages, and supply disruptions caused by unforeseen circumstances, contributing to more resilient, efficient, and sustainable supply chain networks.

For example, a pilot project initiated in cobalt mining in the Democratic Republic of Congo (DRC), where numerous issues related to human rights and environmental concerns were identified, highlighted the advantages technologies can offer (Lewis, 2018; Mugurusi, 2022; IBM, 2019). The implementation of blockchain helped address challenges related to transparency, sourcing, and efficiency. Once the material is mined, a unique identifier is assigned and registered on the blockchain platform, initiating a traceable journey throughout the supply chain. This measure improved transparency by offering traceability from mine to end user, ensuring the authenticity of ethically sourced minerals. It also increased efficiency by reducing supply chain delays by 25% through better coordination and real-time tracking, while administrative costs were cut by 20% through streamlined operations. Additionally, it strengthened trust and confidence in the process, enhanced regulatory compliance by reducing the time and resources needed for reporting by 40%, and bolstered supply chain resilience through real-time tracking and data analysis. Overall, implementing blockchain has demonstrated the potential to significantly improve operations and enhance efficiency.

Similarly, Vestas, a global leader in wind turbine manufacturing, has incorporated AI to optimize supply chain operations for wind energy projects (Choney, 2022; Vestas, 2019). AI has helped the company improve demand forecasting, monitor the availability of critical components, and optimize logistics to ensure the timely delivery and installation of turbines. These AI-driven improvements have reduced delays and material waste that might otherwise occur in the supply chain. Moreover, Vestas uses AI to efficiently identify optimal installation sites based on logistical, environmental, and technical factors, thereby improving overall output.

Technological innovations drive renewable energy, but there is a notable technological and innovation discrepancy between resource-rich and technology-rich countries (Wen et al., 2022). Chile, for instance, has some of the largest reserves of renewable energy materials but lacks the technology to efficiently process those materials domestically (UNCTAD, 2023). As a result, these materials are often transported to foreign countries for processing. This creates dependency on external processing capabilities, making the local renewable energy supply chain vulnerable to disruptions and inefficiencies. Integrating digital technologies can help facilitate the smooth transition and flow of global renewable energy sources. There are several ways to address technological discrepancies in renewable energy supply chains. One approach is capacity building through technology transfer, which can be achieved by encouraging partnerships and joint ventures and implementing training and education programs (Zhang & Gallagher, 2016). These efforts can help develop local capacity and enhance a country's technical skills, reducing dependence on external expertise (Pandey et al., 2023).

One of the many challenges that resource-rich countries face is the lack of adequately educated and trained human resources to manage and operate renewable energy technologies, which also pose significant challenges across the entire value chain. These gaps in skills leave supply chain systems vulnerable to internal and external challenges, deteriorating overall operations. Investments in education and training programs can not only create employment opportunities but also protect the supply chain from disruption, theft, and misuse while building supply chain resilience. Moreover, direct investment by technology-rich countries in resource-rich countries through funding, public–private partnerships, and establishing research and development centers is essential (Ankrah & Lin, 2020). Collaborating on renewable energy projects through aid and alliances with private and public organizations can build renewable energy capabilities in both technologically rich and resource-rich countries (Kruckenberg, 2015).

6.2.3 Logistics and Infrastructure Challenges

Logistics and infrastructure challenges are associated with transportation, distribution, and management of resources essential for renewable energy supply chains. These challenges are often exacerbated by underdeveloped infrastructure and logistical networks, which negatively influence the efficient movement of renewable energy technologies.

The increasing demand for renewable energy technologies has heightened the vulnerability of supply chain networks to unforeseen events, including logistics and infrastructure-related disruptions. In many resource-rich countries, inadequate infrastructure, such as limited road and rail networks and insufficient access to ports or logistical routes, poses significant barriers to the effective extraction and distribution of essential raw materials. This underdevelopment complicates the establishment of resource-based linkages between technologically advanced and resource-rich countries, impacting collaborative efforts in RET production (Gabriel et al., 2016).

Moreover, the fragmented nature of cross-border supply chains further complicates renewable energy operations (UNECA, 2011). Raw materials are often located in remote areas, far from urban centers, creating logistical challenges for their extraction, transportation, and distribution. For instance, in the Democratic Republic of Congo, the sourcing of cobalt is often significantly impacted by the lack of adequate transportation infrastructure, causing delays in shipment. These logistical challenges can lead to inefficiencies, such as prolonged mining project lead times. During periods of high demand for renewable energy technologies, such disruptions can result in supply shortages, exacerbating the situation. This scarcity of essential materials can trigger price hikes, causing manufacturing delays and placing an additional burden on companies to maintain their competitiveness. Consequently, production companies may be compelled to make additional investments, which can disrupt the entire supply chain cycle.

In this context, optimizing logistics through enhanced route planning and infrastructure development can help control costs and minimize the carbon footprint associated with these activities, ensuring the overall efficiency of the supply chain. Improved logistical operations can enhance operational performance and contribute to the industry's overall advancement, leading to a more sustainable and economically viable energy landscape. Technologically advanced countries are typically better positioned to invest in transportation infrastructure. They could contribute to building infrastructure in resource-rich countries, such as roads, high-speed rail, and routes to extraction sites. This requires significant investments in infrastructure development, including ports, railways, and highways, to facilitate the transportation of large components and materials. Such investments can lead to reduced transit time, minimized transportation costs, improved material safety, and overall efficiency in the supply chain. Likewise, fostering international cooperation on shared infrastructure projects can also improve long-distance material transportation efficiency. For example, investment in port facilities in Morocco has significantly enhanced the logistics infrastructure, facilitating the transportation of solar energy technologies. Moreover, countries should diversify their supply chains to minimize dependence on a single supplier while improving production capacities to mitigate potential logistical challenges (UNIDO, 2015).

6.2.4 Geopolitical Challenges

Geopolitical challenges refer to the complex and multifaceted interplay of social, economic, and political dynamics resulting from the interaction between countries and regions, often driven by controlling and distributing resources or pursuing dominance and control. In the context of renewable energy, these challenges include trade wars, international conflicts, local, regional, or international tensions, political instability, and uneven distribution of energy resources across geographic locations. These issues significantly impact the sustainability of supply chains, energy security, and a country's approach to energy production and consumption patterns as they pursue sustainability transitions.

Unlike fossil fuels, concentrated in specific regions and transported through established routes, the renewable energy supply chain is often scattered across multiple countries and continents. While renewable sources like wind and solar energy can be deployed almost anywhere, the materials and minerals necessary are unevenly distributed, complicating the sourcing and distribution capabilities of the actors involved (ERM, 2023). The concentration of these resources in politically sensitive regions and the need to transport them through unstable areas increase supply chain vulnerability. Geopolitical risks such as deglobalization, wars, trade rivalries, ethnocentrism, polarization, and political instability exacerbate these vulnerabilities. Escalating political and ethnic tensions increases the likelihood that powerful groups and cartels will control critical materials. Emerging political and trade tensions, as well as threats like terrorist attacks and military conflicts, negatively influence strategic sourcing and distribution decisions in the renewable energy sector (Riti et al., 2022). Since essential materials are in different countries, political developments within these nations significantly impact the global supply chain, raising concerns regarding energy security, especially rare metals used in manufacturing renewable energy technologies (Su et al., 2021).

Another significant geopolitical challenge in today's world is growing competition for critical raw materials such as lithium, cobalt, and rare earth elements, which are essential for the development of RETs. Global powers, like China and the United States, have been increasingly asserting political pressure on resource-rich countries to secure these supplies, often using trade restrictions on opposing parties to attain political objectives. The emergence of new trade alliances and recent geopolitical developments has also resulted in disruptions to the renewable energy supply chain. For example, the war in Ukraine has disrupted the container routes in the Black Sea, while the continuous threat of attacks from the Houthi rebels has affected logistical movements in the Red Sea and the Gulf of Aden region. Additionally, the sanctions imposed on Russia have triggered price increases and supply shortages, further complicating the supply chains.

Additionally, the act of a country or nation's block can also lead to issues. For instance, military exercises by China in the Taiwan Strait have created uncertainty as stakeholders fear that a major conflict could devastate global supply chains, severely impacting the flow of renewable energy materials and equipment. These geopolitical disruptions can significantly impact the renewable energy sector by increasing prices, delaying delivery, and impacting project outcomes. For instance, shortages of minerals like lithium, cobalt, and other rare earth elements drive up market prices, hampering production capacity and creating a ripple effect throughout the supply chain.

In addition to regional or global challenges, renewable energy supply chains are also vulnerable to the influence of domestic policies within the resource-owning countries. For example, restrictions on exports or nationalization of resources by resource-owing countries can create fear and uncertainty for importing nations. For instance, Indonesia's decision to ban nickel exports to assert sovereignty over its national resources alarmed global manufacturers relying on stable nickel supplies for production. These geopolitical challenges add complexity and instability to global renewable energy supply chains, undermining international efforts to achieve energy transitions and meet climate change goals (IEA, 2022). Such descriptions erode the confidence of industry players and commercial actors, making it crucial to develop resilient supply chains that can withstand geopolitical uncertainties.

To minimize geopolitical risks in supply chain networks, actors and stakeholders can implement several strategies. For example, building resilient supply chains through diversification is essential. Technology-rich countries need to source critical resources from multiple suppliers to reduce over-reliance on a single source. Establishing stable trade and political relations that can endure growing geopolitical volatility is key to ensuring a consistent supply of renewable energy materials. For instance, initiatives like Global Battery Alliance, which promotes sustainable battery supply chains through collaboration among different businesses, governments, and non-governmental organizations, highlight the benefits of cooperative action (GBA, 2024).

Additionally, international collaboration and diplomacy are also central to promoting stability. Bilateral and multilateral agreements that ensure fair practices and protect shared interests can help maintain continuity. In addition, the governments of resource-rich countries must ensure the safe and responsible extraction of resources. Transparency, accountability, and sustainability initiatives can reduce risks associated with resource extraction and minimize negative impacts on the supply chain. To sum up, addressing geopolitical challenges can play an important role in devising resilient, agile, and supply chain networks.

6.2.5 Social and Environmental Challenges

Critical minerals and materials are concentrated in a few places, but their demand is increasing in every part of the world. The supply and demand of these materials radically shape the economic, environmental, and social dynamics of many countries. Each country's environmental and social characteristics are different, which can either hinder or support the flow of resources in the supply chain. While the growth of the renewable energy sector may economically benefit technologically rich countries, logistical activities could pose social and environmental challenges for resource-rich countries. Therefore, achieving efficient, sustainable supply chain management requires considerable focus on the environmental and social impacts of supply chain activities (Cucchiella & D'Adamo, 2013). The logistics and mobility of renewable resources have several environmental impacts (such as greenhouse gas emissions and habitat disruption) and social impacts (such as affecting local communities and changes in land use) (Zang et al., 2020; Abbasi & Nilsson, 2016). Therefore, the identification, monitoring, and addressing of potential negative impacts and consequences along the renewable energy materials supply chain are significant challenges for companies and policymakers (Kügerl et al., 2023). Energy networks can be affected by community concerns and resistance over the aesthetic intrusion of renewable energy technology deployment and the loss of jobs due to the decline of the domestic use of fossil fuel resources (Stigka et al., 2014). Similarly, countries and regional economies dependent on fossil fuel industry profits can hinder supply chain transition policies (Spencer et al., 2017). Moreover, environmental factors, such as preserving natural resources (Paravantis et al., 2014), public concerns over the disruption of the seascape (Seetharaman et al., 2019), and opposition from environmental groups, also impede development (Nasirov et al., 2015).

In developing energy supply networks, there is a need to simplify energy supply chains by focusing on increased transmission capacity, network resilience and security, and reducing the energy supply chain's complexity (Obrecht et al., 2020). Smart mining initiatives, such as sustainably managing mineral resources while fostering economic growth and development, can help resource-rich countries benefit from their resources. At the same time, technologically rich countries can contribute by scaling up their technical assistance and investments in resource-rich countries to ensure minimal environmental and social impacts on supply chain activities (World Bank, 2019). Accordingly, balancing the environmental and social impacts of supply chain activities is crucial for ensuring a sustainable transition to renewable energy.

6.2.6 *Financial Challenges*

Renewable energy offers long-term and cost-effective advantages. Although the renewable energy sector has gained traction among investors, new sustainable energy initiatives require substantial financial investments. Financial challenges are significant in renewable energy chain networks' investment, development, and operational activities (Fernando et al., 2018). However, difficulties in securing initial financing, high costs of initiating projects, market price instabilities, and increased operational costs impact the profitability and planning of activities involved in supply chain networks (Cheng et al., 2023; Infield & Freris, 2020). COVID-19 also severely affected the flow and stability of renewable energy materials; consequently, renewable energy supply chain networks are facing a lack of financing. For example, the financial support given by governments worldwide has been substantially reduced due to the severe effects of the COVID-19 pandemic on their economies (Tsao Thanh and Wei, 2021). Geopolitical situations, such as the current Russia-Ukraine war, the war in Israel, and US-China trade tensions, have also disrupted global renewable energy supply chain networks, increasing the prices for materials needed to create renewable energy technologies, including sharp fluctuations in energy prices worldwide (Cui, 2023). Furthermore, another critical challenge is the low attraction of investors in the clean energy sector due to the high environmental impact and less favorable financial perspectives of renewable energy projects (UNCTAD, 2023).

Addressing financial challenges is fundamental to mitigating risk and building a more resilient renewable energy network. For instance, to secure initial financing, financial burdens, and risks can be minimized through public–private partnerships. Initial costs can be reduced with cost-sharing arrangements among stakeholders, such as a partnership between the resource-rich government and a private renewable energy solar farm in a technology-rich country. The government might offer land, tax incentives, and low-interest loans for the project. This arrangement will lower the company's financial burden and encourage investment in renewable energy projects.

Another example is cost-sharing between stakeholders, such as jointly investing in renewable energy infrastructure to build a grid accommodating multiple renewable energy sources. For instance, this model can work in Africa by constructing hydroelectric dams with the help of international financial institutions and private sector entities to share the costs. Moreover, companies from technologically rich countries could purchase political risk insurance to mitigate the risk of government actions in politically unstable countries that could disrupt renewable energy projects. Market price and geopolitical instabilities can be tackled by making long-term contracts, diversifying sourcing, and using financial instruments against price fluctuations. Additionally, operational costs can be reduced using advanced technology and investing in human resources. Lastly, low investor interest can be addressed by presenting compelling returns on investment models and reducing the social and environmental impacts of renewable energy supply chain activities.

6.2.7 Policy and Regulatory Challenges

A solid legislative framework is required to facilitate the transition to sustainable energy and maintain increased investment in renewable energy (Nathwani & Ng, 2010). Accordingly, developing policies and regulations with lasting and positive implications requires broader engagement from multiple stakeholders to establish and consolidate renewable energy development baselines. Policy and regulatory challenges significantly impact the development and implementation of renewable energy supply chain networks (Yahyazadeh, 2022). For example, the successful development, implementation, and completion of renewable projects require a supportive policy and regulatory framework (Enerdatics, 2023). Research shows that the regulatory environment, natural resources, trade agreements, and social and environmental policies can significantly impact the import and export of energy products (Górecka et al., 2022). Regulatory inconsistencies across different regions create complexity and uncertainty for businesses, hindering the smooth flow of materials between resource-rich and technology-rich countries. Frequent policy changes also create an unstable environment for renewable energy projects. Additionally, the lack of standardization for materials, technologies, human resources, and practices can significantly impact the resource flow from extraction to usage. Trade and tariff barriers to the supply of materials are also significant challenges for companies, affecting costs and limiting access to essential materials required to produce renewable technologies. Social and environmental impact regulations can also hinder renewable energy supply networks (Abate et al., 2016). For instance, resource extraction may severely impact the environment and nearby communities. Accordingly, heightened awareness of the negative environmental and social impacts of resource depletion has led to stringent regulatory policies in many countries (UCSUSA, 2017).

The policy and regulatory challenges can be tackled with various steps to ensure a smooth flow of materials and cooperation in regulatory matters. For instance, regulatory inconsistencies can be addressed by establishing bilateral agreements, creating regulatory liaisons, and harmonizing regulatory policies. Policymakers in both resource-rich and technology-rich countries should work on policy stability by engaging different stakeholders and monitoring the impact of policy changes on supply chain networks. Trade and tariff barriers on materials can be tackled through agreements and the provision of incentives and subsidies to resource-rich countries, supported by technology-rich countries, to bolster the renewable energy sector. Most importantly, the social and environmental impacts of resource extraction and production can be managed by following sustainable practices and codes, engaging with communities, and providing compliance support.

6.2.8 Governmental Challenges

Uneven government subsidies and support remain one of the biggest challenges for companies developing renewable energy supply chains (Fernando et al., 2018). Although numerous governments globally are encouraging businesses to switch to renewable energy with tax advantages, investments, subsidies, and other favorable measures, this support is not consistent across all regions. Due to the Russia-Ukraine war and the recovery from the COVID-19 pandemic, governments have supported businesses and consumers, leading to increased fossil-fuel subsidies worldwide. The International Monetary Fund (IMF) has reported that to decrease fuel prices, both implicit and explicit fossil-fuel subsidies have surged to a record 7 trillion dollars, especially in developing countries with higher pollution levels generated by factories and vehicles (IMF, 2023).

In some instances, governments favor fossil fuel industries over renewable energy companies under the guise of environmental and community protection, complicating the situation and making compliance with such regulations challenging for new entrants. Developed countries' governments often favor the fossil fuel industry because they enjoy sizable revenue flows from fossil fuels in terms of royalties and taxes. Inflexible tender designs to procure and contract raw materials further add cost pressure along the supply chain (Fischer et al., 2023). The lack of qualified human resources, complex authorization procedures for renewable energy projects, and poor coordination among governmental agencies further complicate the smooth flow of renewable energy resources between resource-rich and technology-rich countries. This means that renewables are not entering the market conveniently, competing with fossil fuel industries, which are favored and encouraged by subsidies and incentives, while polluters are not being penalized (UCSUSA, 2017).

Governments can play a crucial role in transitioning from fossil fuels to renewable energy sources. For example, together with technology-rich country governments, resource-rich governmental organizations can create public awareness and drive political change. They can agree on common goals and objectives to devise policies that help businesses streamline supply chain networks. To reduce the lengthy process of project approval, there should be clear guidelines, decentralization, and the use of digital technologies to streamline the process. In developing countries, poverty is one of the biggest causes and effects of environmental problems. Therefore, technology-rich countries can help resource-rich countries separate environmental development issues from economic development issues. Resource-rich governments should be incentivized and assisted in policy reforms by technology-rich countries to remove biases favoring the fossil fuel industry. To enhance the resilience of renewable energy supply networks, it is important that governments reduce subsidies to fossil fuel industries and encourage the renewable energy industry, resulting in improved supply and increased energy efficiency.

6.3 Conclusion and Future Trends

The unprecedented increase in population worldwide necessitates a significant increase in energy demand. For many years, countries have relied on fossil fuels for consumption and industrial energy needs. Although energy is crucial for economic and human advancement, using fossil fuels contributes to global warming by releasing carbon dioxide (CO_2) and greenhouse gases (GHG) into the atmosphere, thereby exacerbating climate change issues. Alternatively, renewable energy sources reduce global warming without degrading the ecosystem. However, given the increased demand for renewable energy technologies, the global renewable energy market is affected by intermittent resource availability, leading to production and installation delays, power disruptions, consumer distrust and discouragement, and decreased energy efficiency. Supply chains play a central role in the renewable energy transition by delivering the technology's essential materials and minerals. However, many challenges may stall this transition. Supply chain disruptions not only hurt renewable energy companies but also slow down the transition to renewable energy. Therefore, there is a pressing need for sustainable, efficient, and secure supply chain networks to streamline the flow of renewable energy resources.

The findings of this study may guide policymakers, companies, and developers in understanding the range of supply chain challenges hindering progress toward renewable energy in both resource-rich and technology-rich countries. For example, geographical advantages and disadvantages are becoming major limitations impacting the development and deployment of renewable energy. Geographical challenges can be addressed by strengthening bilateral relationships between countries to facilitate and manage resources, supporting political stability, and developing infrastructure. Similarly, technological challenges can be addressed through capacity building, technology transfer, investments in education and training programs, and the establishment of research and development centers through private–public partnerships. This chapter further identifies that logistical challenges significantly hinder the efficiency of the entire supply chain network. In this situation, the resilience of supply chain networks can be enhanced through infrastructure development and searching for alternate suppliers. Likewise, geopolitical risks can be managed by diversifying energy supply away from politically unstable countries, establishing stable trade and political relations, and ensuring the safe and responsible extraction, transmission, and distribution of resources through accountability, transparency, and sustainability efforts.

To avoid social and environmental risks, corporations may implement smart mining initiatives to sustainably manage materials while supporting the development and growth of the local economy. Scaling up investments and providing technical assistance may ensure minimal social and environmental impact from supply chain activities. Additionally, with the help of research and development initiatives, operational costs can be reduced, the efficiency of supply chain networks can be enhanced, and social and environmental risks can be mitigated. The importance of financial challenges in the smooth functioning of supply chain networks cannot be underrated. Renewable companies are heavily dependent on critical materials, and the increase in the cost of even a single item can hinder the progress of entire projects. In this situation, firms cannot bear the financial burden of rising raw material prices due to inflation and persistent disruptions in supply chains. The financial challenges in supply chains can be minimized through public–private partnerships, using financial instruments against price fluctuations, and addressing investors' concerns by presenting compelling investment returns.

To address policy and regulatory challenges, decision-makers in resource-rich and technology-rich countries need to collaborate on streamlining their policy and regulatory mechanisms. For example, when devising policies and making regulations, monitoring their impact on renewable energy projects, and considering different stakeholders should be kept in mind. Most importantly, drafting supportive social and environmental policies, such as following sustainable codes, engaging with communities, and providing compliance support, can reduce issues in supply chain networks. Similarly, the lack of public awareness about the benefits of renewable energy and the governmental will to continue using non-renewable energy sources can pose significant challenges to the success of renewable energy projects. Creating public awareness and driving political change can be achieved through mutual collaboration, agreements on common goals and objectives, and pushing governments to reduce reliance on fossil fuels. Additionally, supply chain challenges can be minimized by strategies such as providing a supportive environment to renewable energy companies, incentivizing them through tax incentives, easing trade barriers, and implementing renewable energy-friendly measures to facilitate the transition toward renewable energy.

Rapid growth and dependency on renewable energy resources would further require resilient supply chain networks. Accordingly, using emerging technologies requires addressing future challenges to improve efficiency, reduce operational costs, provide a transparent and secure system, and improve scalability (Taherdoost, 2024). For example, using blockchain in the supply chain would improve transparency, security, and trust among stakeholders. At the same time, using artificial intelligence and data analytics can help predict disruptions, price fluctuations, human resource management, and enhance material flow, including helping manage material and overall project management. Furthermore, the use of 3D printing technology in the manufacturing of the components of renewable energy technologies can save costs and ensure the availability of needed parts. In addition, to effectively manage and quickly respond to disruptions in the supply chain, companies can use the Internet of Things (IoT). Digital Twin Technology can also help companies simulate supply chain activities to identify bottlenecks, thus optimizing the whole supply chain. Supply chain network challenges in the future can be effectively managed with the help of robotics and automation in material extraction and manufacturing. (See Fig. 6.3).

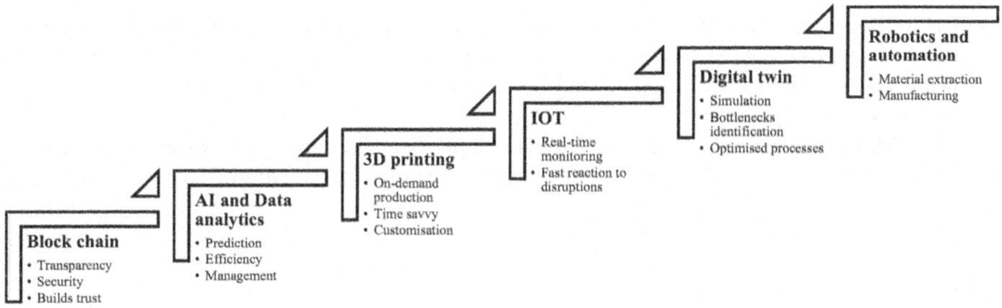

Fig. 6.3 Future trends to mitigate challenges in renewable energy networks (Authors' work)

Acknowledgments Shah Rukh Shakeel acknowledges the funding from the Research Council of Finland (funding decision number 366010) and the Foundation for Economic Education (Grant number 180325).

References

Abate, G., Nielsen, M., & Tveterås, R. (2016). Innovation and renewable energy markets: The role of policies and market structure. *Energy Policy, 94*, 156–166.

Abbasi, M., & Nilsson, F. (2016). Developing environmentally sustainable logistics: Exploring themes and challenges from a logistics service providers' perspective. *Transportation Research Part d: Transport and Environment, 46*, 273–283.

Almutairi, K., Hosseini Dehshiri, S. J., Hosseini Dehshiri, S. S., Hoa, A. X., Arockia Dhanraj, J., Mostafaeipour, A., & Techato, K. (2023). Blockchain technology application challenges in renewable energy supply chain management. *Environmental Science and Pollution Research, 30*(28), 72041–72058.

Ankrah, I., & Lin, B. (2020). Renewable energy development in Ghana: Beyond potentials and commitment. *Energy, 198*, 117356.

Bridge, G., Bouzarovski, S., Bradshaw, M., & Eyre, N. (2013). Geographies of energy transition: Space, place and the low-carbon economy. *Energy Policy, 53*, 331–340.

Cheng, J., Mohammed, K. S., Misra, P., Tedeschi, M., & Ma, X. (2023). Role of green technologies, climate uncertainties and energy prices on the supply chain: Policy-based analysis through the lens of sustainable development. *Technological Forecasting and Social Change, 194*, 122705.

Choney, S. (2022). How one of the world's largest wind companies is using AI to capture more energy. Microsoft News. https://news.microsoft.com/source/emea/features/winds-of-change-how-one-of-the-worlds-largest-wind-companies-is-using-ai-to-capture-more-energy/. Accessed 15 Oct 2024.

Clean energy minerals: Developing countries must add value to Capitalise on demand. (2023). United Nations Trade and Development Report. https://unctad.org/news/clean-energy-minerals-developing-countries-must-add-value-capitalize-demand. Accessed 16 July 2024.

Cucchiella, F., & D'Adamo, I. (2013). Issue on supply chain of renewable energy. *Energy Conversion and Management, 76*, 774–780.

Cui, Y. (2023). Geopolitical risks and renewable energy supply chains: Implications and resilience strategies. *Journal of Energy Policy and Strategy, 48*(2), 87–101.

Enerdatics. (2023). *Trends in renewable energy project development: Market dynamics and policy frameworks*. Enerdatics Report. Available at https://www.enerdatics.com/. Accessed 12th May 2024.

ERM (Environmental Resources Management). (2023). *The future of renewable energy supply chains: Managing geopolitical risks*. ERM Insights Report. https://www.erm.com/insights/renewables-conundrums---four-steps-to-bolster-renewable-energy-supply-chains/. Accessed 17th July 2024.

Feng, L., Zhang, X., & Chen, S. (2024). Global competition and cooperation in renewable energy materials: Challenges and opportunities. *Renewable and Sustainable Energy Reviews, 180*, 113187.

Feng, Y., Hu, J., Afshan, S., Irfan, M., Hu, M., & Abbas, S. (2023). Bridging resource disparities for sustainable development: A comparative analysis of resource-rich and resource-scarce countries. *Resources Policy, 85*, 103981.

Fernando, Y., Bee, P. S., Jabbour, C. J. C., & Thomé, A. M. T. (2018). Understanding the effects of energy management practices on renewable energy supply chains: Implications for energy policy in emerging economies. *Energy Policy, 118*, 418–428.

Fischer, L. B., Frankl, P., & Bahar, H. (2023). Managing critical materials in energy transitions: Policy tools for resilient supply chains. *Energy Policy, 178*, 113559.

Gabriel, C. A. (2016). What is challenging renewable energy entrepreneurs in developing countries? *Renewable and Sustainable Energy Reviews, 64*, 362–371.

GBA (Global Battery Alliance). (2024). Establishing a sustainable and responsible battery value chain. https://www.globalbattery.org/. Accessed 15 Oct 2024.

Górecka, D., Skender, H., & Zaninović, V. (2022). Policy influences on the international trade of renewable energy products: A cross-country analysis. *Renewable Energy, 187*, 1080–1092.

IBM. (2019). Blockchain for the mining industry: Ethical cobalt production. https://www.ibm.com/blog/blockchain-for-the-mining-industry-ethical-cobalt-production/ Accessed 30 July 2024.

IEA. (2024). World Energy Investment 2024. https://www.iea.org/reports/world-energy-investment-2024. Accessed 18 Oct 2024.

IMF (International Monetary Fund). (2023). Fossil fuel subsidies surged to record $7 trillion: Scaling back subsidies would reduce air pollution, generate revenue, and make a major contribution to slowing climate change. https://www.imf.org/en/Blogs/Articles/2023/08/24/fossil-fuel-subsidies-surged-to-record-7-trillion. Accessed 5 August 2024.

International Energy Agency (IEA). (2022). *Global supply chains of critical minerals for clean energy technologies*. Paris: IEA. https://www.iea.org/topics/critical-minerals. Accessed 19th April 2024.

International Energy Agency (IEA). (2024). Renewables 2023-Analysis and forecast to 2028. https://iea.blob.core.windows.net/assets/96d66a8b-d502-476b-ba94-54ffda84cf72/Renewables_2023.pdf. Accessed 18 Oct 2024.

International Renewable Energy Agency (IRENA). (2021). *World energy transitions outlook: 1.5°C pathway*. Abu Dhabi: IRENA. https://www.irena.org/-/media/Files/IRENA/Agency/Publication/2021/March/IRENA_World_Energy_Transitions_Outlook_2021.pdf. Accessed March 15th, 2024.

International Renewable Energy Agency (IRENA). (2024). *Renewable capacity statistics 2024*. Abu Dhabi: IRENA. https://www.irena.org/Publications/2024/Mar/Renewable-capacity-statistics-2024. Accessed 21 April 2024.

IPCC (Intergovernmental Panel on Climate Change). (2018). *Global warming of 1.5°C: An IPCC special report*. Geneva: IPCC. https://www.ipcc.ch/sr15/2018/. Accessed 20 January 2024

Infield, D., & Freris, L. (2020). *Renewable energy in power systems*. Wiley & Sons.

Kruckenberg, L. J. (2015). North-South partnerships for sustainable energy: Knowledge–power relations in development assistance for renewable energy. *Energy for Sustainable Development, 29*, 91–99.

Kügerl, M., Hofmann, L., & Stigler, H. (2023). Digitalization of energy systems: Opportunities and challenges for renewable energy integration. *Renewable Energy Focus, 45*, 24–31.

Lewis, B. (2018). Blockchain to track Congo's cobalt from mine to mobile. Reuters. https://www.reuters.com/article/business/finance/blockchain-to-track-congos-cobalt-from-mine-to-mobile-idUSKBN1FM0Y2/ .Accessed 20 July 2024.

Li, N., Dilanchiev, A., & Mustafa, G. (2023). From oil and mineral extraction to renewable energy: Analysing the efficiency of green technology innovation in the transformation of the oil and gas sector in the extractive industry. *Resources Policy, 86*, 104080.

Mugurusi, G., & Ahishakiye, E. (2022). Blockchain technology needs for sustainable mineral supply chains: A framework for responsible sourcing of Cobalt. *Procedia Computer Science, 200*, 638–647.

Nasirov, S., Silva, C., & Agostini, C. A. (2015). Investors' perspectives on barriers to the deployment of renewable energy sources in Chile. *Energies, 8*, 3794–3814.

Nathwani, J., & Ng, A. W. (2010). *Backcasting a future of sustainable energy: A public policy perspective*. INTECH Open Access Publisher.

Obrecht, M., Kazancoglu, Y., & Denac, M. (2020). Transition towards sustainable energy systems: A supply chain perspective. *Renewable and Sustainable Energy Reviews, 133*, 110200.

Pandey, V., Sircar, A., Bist, N., Solanki, K., & Yadav, K. (2023). Accelerating the renewable energy sector through industry 4.0: Optimisation opportunities in the digital revolution. *International Journal of Innovation Studies, 7*(2), 171–188.

Paravantis, J., Stigka, E., & Mihalakakou, G. (2014). An analysis of public attitudes towards renewable energy in Western Greece. *Renewable and Sustainable Energy Reviews, 32*, 100–106.

Paris Agreement. (2024). *Paris climate accord: Updated commitments and progress report*. United Nations Framework Convention on Climate Change (UNFCCC). https://legal.un.org/avl/pdf/ha/pa/pa_e.pdf. Accessed 20 March 2024

REN 21. (2024). Renewable 2024. Global Status Report. Global Overview. https://www.ren21.net/gsr-2024/modules/global_overview/02_policy/. Accessed 18 October 2024

REN (21). (2021). Renewables 20243. Global Status Report Collection, Renewables in Energy Demand. https://www.ren21.net/gsr-2023/modules/energy_demand/. Accessed 15 July 2024.

Riti, J. S., Shu, Y., & Riti, M. K. J. (2022). Geopolitical risk and environmental degradation in BRICS: Aggregation bias and policy inference. *Energy Policy, 166*, 113010.

Samadi, S., Fischer, A., & Lechtenböhmer, S. (2023). The renewables pull effect: How regional differences in renewable energy costs could influence where industrial production is located in the future. *Energy Research and Social Science, 104*, 103257.

Sánchez-García, J., Cortés-García, F. J., & Gualteros Sánchez, L. F. (2024). Smart technologies and renewable energy supply chains: Challenges and future research directions. *Renewable and Sustainable Energy Reviews, 184*, 113392.

Sarkis, J., Kouhizadeh, M., & Zhu, Q. S. (2021). Digitalisation and the greening of supply chains. *Industrial Management and Data Systems, 121*(1), 65–85.

Seetharaman, A., Kumar, N., Palaniappan, S., & Weber, G. (2019). Factors affecting renewable energy adoption. *Renewable and Sustainable Energy Reviews, 107*, 555–564.

Shakeel, S. R., Takala, J., & Zhu, L. D. (2017). Commercialisation of renewable energy technologies: A ladder building approach. *Renewable and Sustainable Energy Reviews, 78*, 855–867.

Spencer, T., Berghmans, N., Sartor, O., & Pierfederici, R. (2017). The role of international trade in achieving sustainable energy transitions. *Climate Policy, 17*(8), 973–989.

Statista. (2024). New investment in renewable energy worldwide from 2014 to 2023. https://www.statista.com/statistics/186807/worldwide-investment-in-sustainable-energy-since-2004/ Accessed 18 October 2024.

Stigka, E. K., Paravantis, J. A., & Mihalakakou, G. K. (2014). Social acceptance of renewable energy sources: A review of contingent valuation applications. *Renewable and Sustainable Energy Reviews, 32*, 100–106.

Su, Z. W., Umar, M., Kirikkaleli, D., & Adebayo, T. S. (2021). Role of political risk to achieve carbon neutrality: Evidence from Brazil. *Journal of Environmental Management, 298*, 113463.

Taherdoost, H. (2024). Blockchain integration and its impact on renewable energy. *Computers, 13*(4), 107.

Tsao, Y. C., Thanh, V. V., Lu, J. C., & Wei, H. H. (2021). A risk-sharing-based resilient renewable energy supply network model under the COVID-19 pandemic. *Sustainable Production and Consumption, 25*, 484–498.

UCSUSA (Union of Concerned Scientists). (2017). *The hidden costs of fossil fuels*. Cambridge, MA: Union of Concerned Scientists. https://www.ucsusa.org/resources/barriers-renewable-energy-technologies. Accessed 3rd August 2024.
UNCTAD. (2023). United Nations Trade and Development Report. https://unctad.org/news/clean-energy-minerals-developing-countries-must-add-value-capitalize-demand. Accessed 16th July 2024.
UNECA (United Nations Economic Commission for Africa). (2011). *Renewable energy in Africa: Prospects and limits*. Addis Ababa: UNECA. https://repository.uneca.org/bitstream/handle/10855/21569/Bib-69220.pdf?sequence=1. Accessed 16th July 2024.
UNIDO (United Nations Industrial Development Organization). (2015). *Promoting renewable energy investments: Case studies and lessons learned*. Vienna: UNIDO. https://downloads.unido.org/ot/99/28/9928078/WP21.pdf. Accessed 16th July 2024.
United Nations (UN). (2024a). *Science, solutions, solidarity for a livable planet*. https://www.un.org/en/climatechange. Accessed July 10th, 2024.
United Nations (UN). (2024b). *Sustainable energy for all progress report 2024*. New York: United Nations. https://digitallibrary.un.org/record/4051770?v=pdf. Accessed July 10th, 2024.
VESTAS. (2019). Leading the way with renewable solutions. https://www.vestas.com/content/dam/vestas-com/global/en/investor/reports-and-presentations/presentations/capital-markets-day-2021/Part%20III.pdf. Accessed 15 Oct 2024.
Wen, F., Liao, H., & Liu, Y. (2022). Technology gaps and innovation barriers for renewable energy development: Evidence from emerging economies. *Energy Strategy Reviews, 41*, 100840.
World Bank. (2019). Climate-smart mining: Minerals for climate action. https://www.worldbank.org/en/topic/extractiveindustries/brief/climate-smart-mining-minerls-for-climate-action. Accessed 23 July 2024.
World Future Council. (2024). *Shaping the renewable energy future: Global policy innovations*. Hamburg: World Future Council. https://www.worldfuturecouncil.org/global-renewable-energy-goal/. Accessed July 10th, 2024.
World Trade Organization (WTO). (2024). *World Trade Report 2024*. https://www.wita.org/wp-content/uploads/2024/09/WTO-2024-World-Trade-Report.pdf. Accessed 10 March 2024.
Yahyazadeh, A. (2022). Review of renewable energy challenges and opportunities in supply chain. *International Journal of Engineering and Technology Sciences*, 1–25.
Yu, H., Wen, B., Zahidi, I., Chow, M. F., Liang, D., & Madsen, D. Ø. (2024). The critical role of energy transition in addressing climate change at COP28. *Results in Engineering*, 102324.
Zang, G., Liu, Z., & Wang, Y. (2020). Blockchain for sustainable supply chain management: A review and future research directions. *Resources, Conservation and Recycling, 154*, 104596.
Zhang, F., & Gallagher, K. S. (2016). Innovation and technology transfer through global value chains: Evidence from China's PV industry. *Energy Policy, 94*, 191–203.

Chapter 7
Energy Transition and Social Justice: Examining the Impact of Renewable Energy Supply Chain

Tommi Lehtonen and Shah Rukh Shakeel

Abstract The scholarly discussion on renewable energy technology development has primarily focused on technological advancements and their integration into global energy systems. The supply chain's role in carrying out technology development processes has received little attention. Exploring renewable energy development from a life cycle perspective reveals implications for social justice that transcend mere technological development. In terms of social justice, the biggest challenge in the energy transition is fairly distributing benefits and expenses to various groups of people across the renewable energy supply chain. The negative environmental impacts of energy production and consumption are often borne by marginalised communities. Therefore, it is crucial to ensure that the public, understood broadly, has a say in how the transition is planned and implemented. This means ensuring that there is public participation in decision-making processes, that communities are consulted about proposed energy transition projects, and that there is transparency in the decision-making process.

Keywords Energy transition · Social justice · Renewable energy · Supply chain · Energy poverty

The Research Council of Finland funded the research for this chapter—funding decision number 366010.

T. Lehtonen (✉) · S. R. Shakeel
University of Vaasa, Innovation and Entrepreneurship InnoLab, Vaasa, Finland
e-mail: tommi.lehtonen@uwasa.fi

S. R. Shakeel
e-mail: shah.rukh.shakeel@uwasa.fi

I. Rahimi et al. (eds.), *Sustainable Supply Chain of Renewable Energy Networks*, Energy Systems in Electrical Engineering, https://doi.org/10.1007/978-981-96-4940-2_7

7.1 Introduction

The current global polycrisis, including climate change, biodiversity loss, and various types of social inequality, requires behavioural changes from individuals, communities, nations, and corporations. Moving from fossil fuels, including oil, coal, and natural gas, to renewable energy sources, such as solar power, wind, and hydro, is one of the most important steps to counteract biodiversity loss and climate change.

The 2023 United Nations Climate Conference in Dubai reached an agreement to shift away from fossil fuels—a particularly significant decision as it represented the first time a global climate conference resolution has included such a transition. This sends a strong signal to investors, who will now have a more difficult time justifying investments in the fossil economy. As expressed in the 2015 Paris Climate Agreement and confirmed in the Dubai resolution, the ultimate goal is to achieve zero global emissions by 2050 while limiting global warming to 1.5 °C above pre-industrial levels by the end of the century.

While there are valid concerns about funding for low-carbon technology to meet global climate targets, many experts believe that the most difficult problem will be managing the energy transition in a way that does not exacerbate socioeconomic inequities (Sun et al., 2023). Millions of people have been pulled out of poverty by earlier industrial transformations, but these changes have also had the consequence of solidifying the position of power and wealth of the ruling class. There is a high risk that vulnerable populations may be left behind as countries transition to cleaner energy sources.

The transition from centralised energy systems to decentralised, renewable-based sustainable energy systems is a complex and multifaceted process influenced by various factors, including technology, resources, supportive policy regimes, and changes in consumption patterns (Balachandra et al., 2010; Droege et al., 2010; Hrnčić et al., 2021; Papaefthymiou & Dragoon, 2016; Shakeel et al., 2017). While the development of renewable energy technologies (RETs) is critical for a smooth transition, the mechanisms used across the RET supply chain have a significant impact on delivering justice.

To date, the scholarly discussion on energy transition has primarily focused on technological advancements and their integration into global energy systems. Renewable energy and efficient energy storage technologies are viewed as critical to decreasing reliance on traditional hydrocarbons as energy sources (Balachandra et al., 2010; IRENA, 2023; Ould Amrouche et al., 2016; Papaefthymiou & Dragoon, 2016; Shakeel, 2019). The economic and legal frameworks needed to facilitate the just transition and its possible consequences on emissions, the environment, and energy access and affordability have also received attention (García-García et al., 2020; Jenkins et al., 2016, 2021; Lacey-Barnacle, 2020; Malerba, 2022; Sovacool & Dworkin, 2015).

However, the role of the supply chain in technology development processes has received little attention. Exploring renewable energy development through a life cycle perspective reveals implications for social justice that extend beyond technology adoption. Strengthening justice within the renewable energy supply chain is crucial for ensuring that the transition benefits all stakeholders equitably rather than only those directly benefiting from technology application. This holistic approach ensures a consistent commitment to addressing the interests and concerns of all actors and communities through the energy supply chain.

Several examples illustrate how injustices in the renewable energy development process have led to inequities and broader social harm. For example, lithium mining operations in Salar de Uyuni, Bolivia, highlight significant energy justice issues. Lithium, a critical mineral for energy storage and the development of renewable energy technologies, is essential to the global energy transition. However, its extraction has caused several environmental damages, including water depletion that affects local agriculture and drinking supplies. Moreover, ongoing conflicts between local communities and national authorities over revenue distribution, coupled with issues related to transparency and fairness, have exacerbated tensions (Radwin, 2024; Sagárnaga, 2015; Sanchez-Lopez, 2021). Similarly, manufacturing renewable energy technologies, such as solar panels, in countries with lower labour costs perpetuate serious injustices, exposing workers to poor working conditions, low wages, and minimal labour rights (E + ELeader: For a Sustainable Future 2024).

Addressing these issues within the renewable energy sector can help mitigate social disparities, prevent unfair practices, and facilitate a fair transition by implementing inclusive measures, distributing resources fairly, and ensuring transparency and fairness throughout the supply chain. This will ensure that the shift to renewable energy is not only ecologically but also socially sustainable.

This book chapter explores how integrating social justice during the renewable energy supply chain can facilitate a just transition from conventional to sustainable energy systems based on renewables. The findings of this study provide valuable theoretical insights into social justice issues influencing the development of renewable energy technologies. On a practical level, the chapter offers actionable guidelines for companies and policymakers to align RET development with the principles of social justice.

To achieve this, we will identify and discuss significant social justice concerns raised by the energy transition, drawing on conceptual analyses and a comprehensive review of the literature. Recent reports and guidelines from major international agencies, such as the European Union (EU), the United Nations (UN), and the International Labour Organisation (ILO), along with real-life examples, provide particularly useful information about the social and economic challenges associated with the energy transition.

This book chapter is structured as follows: In the next section, we present a brief outline of the energy transition and discuss its historical context. Following that, through the concept of sufficiency, we explore the relationship between social justice and environmental sustainability—together, they form the theoretical lens through which we examine the existing literature on energy transition. In connection with this, we also present the 'doughnut economics' model and discuss a range of issues and challenges related to energy transition, such as energy poverty. Different ethical models are employed to provide a comprehensive analysis of these issues.

We then delve into the concept of energy justice and its implications for the energy sector. Next, we examine energy justice challenges across the renewable energy supply chain, highlighting their significance for a just transition through various cases. Finally, based on these insights, we offer recommendations for policymakers on ensuring justice in energy transition by incorporating a supply chain perspective.

7.2 Basics of Energy Transition

As an important prelude to discussing energy justice, we will now briefly describe the key elements and starting points of the energy transition. Many readers are already familiar with the core aspects of the energy transition but will find it useful to revisit a few of them as an introduction to energy justice.

An energy transition is, by definition, a change in the energy sources required to meet global energy demand (Valkhof, 2020). Using fossil fuels accelerates global warming and endangers biodiversity. This drives the energy transition. The transition's goal is to protect the environment and climate from the harmful impacts of human activities.

The energy transition has many facets. First, it means a gradual move away from fossil fuel-based production and consumption. We are transitioning from oil, coal, and natural gas to green energy sources such as bioenergy, wind, solar, and hydropower. The energy transition also includes regulating energy, electricity storage, and demand response. Regulating energy refers to the part of production needed in electricity generation in addition to baseload power, which is utilised when demand for electricity exceeds baseload power generation capacity. Demand response refers to matching electricity demand to electricity availability, which means limiting electricity use and shifting it from hours of high consumption and prices to more favourable times. The energy transition also includes the shift from centralised to more decentralised generation and the electrification of transport and industry. Energy transition can also refer to efficiency gains and the use of industrial by-products such as processing waste or surplus materials. This ties the circular economy and the energy transition together.

Many nations are already shifting to low-carbon energy sources, driven by economic or health concerns and bolstered by climate change legislation (United Nations, 2021). However, governments, corporations, employees, and communities all tend to 'hasten slowly' and keep carbon-intensive industries alive for employment and economic reasons, as well as for a variety of social reasons.

The energy transition entails transforming energy production as well as accelerating the implementation of low- or zero-carbon technology in industry, transportation, agriculture, and households, to name a few sectors. Early engagement in a just energy transition in all sectors and by various stakeholders, such as investors, employees, and residents, can reduce negative repercussions while increasing positive outcomes. Therefore, the 2015 Paris Climate Agreement highlights just transition as a key principle. The Paris Agreement defines just transition as a vision and a dialogue-based process rather than a defined set of rules. Accordingly, it is a shared agenda among workers, industry, and governments that must be negotiated and implemented within their various geographical, political, cultural, and social settings. For the energy transition to be fair, it must adhere to a set of ethical and social norms, such as the International Labour Organisation's Just Transition Guidelines (ILO, 2015).

Historically, the world has undergone several transitions, including automation and the collapse or relocation of entire industries, resulting in job losses and economic misery. People are concerned that future transitions will be equally difficult. The present energy transition represents the fourth major shift in energy sources (Valkhof, 2020). First, coal replaced wood as a biofuel (1830–1950). The second energy transition (1950–1980) developed and used refined oil products. The third transition (1980–2020) increased natural gas use. As examples of the fourth or current energy transition, wind and solar power have recently exceeded fossil natural gas in energy generation in Europe for the first time (European Commission, 2023). Electric vehicle sales have also increased significantly (IEA, 2023a).

Sustainability considerations, the most important of which is the need to combat climate change, drive the current energy transition. The past transitions were mostly driven by a desire for economic growth, which is closely linked with energy availability and consumption. In the current energy transition, affordability and access to renewable energy technologies are critical issues that ensure that the transition benefits all members of society.

7.3 Social Justice

Social justice is the idea that everyone should be treated fairly, no matter their race, gender, sexual orientation, disability, or social class (Whaples et al., 2023). Today, some of the most important problems in social justice are economic inequality, racial injustice, gender injustice, disability injustice, and environmental injustice. Economic inequality and environmental injustice are the most pressing social justice issues surrounding the energy transition. However, a shortage of fairly priced green energy may reveal unique challenges related to race, gender, and disability that require consideration.

Many, if not all, countries have widening gaps between the rich and the poor (Blanchet & Martínez-Toledano, 2023). People of low-income communities are frequently disproportionately impacted by environmental pollution and climate change (Morello-Frosch & Obasogie, 2023). Women and girls have historically been denied similar rights and opportunities as men, and they face discrimination in employment, education, and politics, as well as potentially limited access to energy resources and new technology (Nations, 2014; Shang Baoping, 2022). Poverty, crime, and violence disproportionately affect people of colour, and they face discrimination in the workplace, housing market, and educational system (Pager & Shepherd, 2008). All of these facets of social injustice must be considered in planning and implementing the energy transition.

A major ethical issue in social justice is whether wealth, such as energy resources, should be distributed as evenly as possible or whether it is sufficient that everyone's basic needs are addressed. Accordingly, there are two extremes to social justice: An equality concept and a sufficiency perspective. As a result, the concept of social justice can be refined and characterised in different ways. This conceptual issue needs to be clarified since it affects the entire subject of social justice.

7.3.1 Sufficiency

Recent discussions in sustainability studies have emphasised the idea of sufficiency, a concept that resonates with our research on energy justice as it addresses both social justice and environmental concerns (Princen, 2003; Salleh, 2009; Kanschik, 2016; Bonnedahl & Heikkurinen, 2019; Lehtonen & Heikkurinen, 2022). More precisely, sufficiency relates to energy justice by tackling the problem of just and adequate renewable energy availability while also highlighting the potential risks associated with excessive and unnecessary energy use.

Aside from these preliminary remarks, the concept of sufficiency must be examined in further depth. On the one hand, 'sufficiency' can refer to the state and characteristic of being adequate or having enough—not too much or too little. (Hartmann, 2024). On the other hand, when a task or resource is sufficient, it indicates it is exactly right and proper. Thus, sufficiency, whether defined as a situation or a quality, represents the 'golden mean' between not having enough and having too much, or between exaggerating and downplaying (Manno, 2017).

Sufficiency also entails balancing, either in quantity or quality, and establishing what is required—for example, how much energy is required under given conditions. When something is either too little or too much, an imbalance results—for instance, between supply and demand for renewable energy. In their 2016 study, U-tantada et al. say that we can think of sufficiency in terms of traits like moderation, equity, and reasonableness. The sufficiency idea thus involves an ethical element by nature (U-tantada et al., 2016).

In terms of sustainability, sufficiency typically refers to limits of production and consumption that, when surpassed, make the use of human-made resources (such as money, tools, equipment, and buildings) as well as natural resources (such as land, water, minerals, plants, and animals) unfair, harmful, or excessive (Princen, 2003). Raworth's (2017) 'doughnut economics', a popular example of the limits model, asserts the existence of a lower limit, the social foundation, below which no one should fall, and an upper ecological limit that one should not cross. Accordingly, the essence of doughnut economics consists of two major boundaries: A social foundation of wellbeing that no one should fall below and an ecological ceiling of planetary pressure that we must not exceed. Between the two is a safe and equitable area for all (Fig. 7.1).

The doughnut economics, then, teaches us that we must weigh everyone's desire for additional resources against everyone else's needs and available resources.

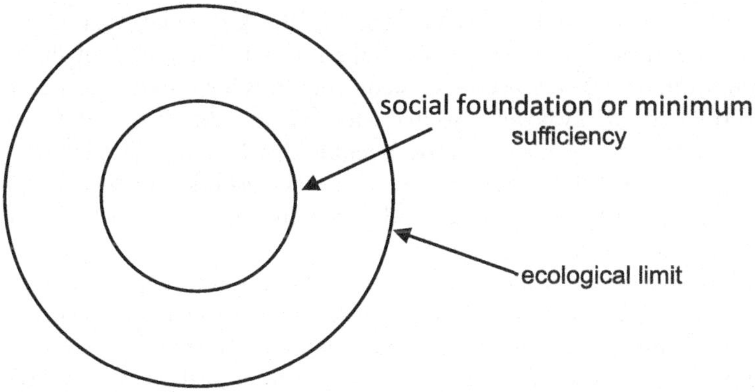

Fig. 7.1 Doughnut economics

Furthermore, a just allocation of resources must adhere to the ecological limit of sustainability.

When all these requirements are considered, significant dimensions of energy justice seem to include at least the following:

(1) Energy availability or the sufficiency of energy supply
(2) Affordability of energy
(3) Availability of affordable green energy (non-fossil).

When these three dimensions of energy justice are examined through the lens of the doughnut economy, we may identify thresholds below which there is insufficient energy and above which the usage of energy exceeds the ecological safety threshold—that is, the planetary boundaries.

When sufficiency is considered as a moral principle, there is the issue of the equal division of energy across all three dimensions. Most, if not all, societies have the energy necessary to meet everyone's demands, but some people cannot afford it. In addition, some can afford green energy, while others must rely on fossil fuels and environmentally unfriendly technology such as fuel-intensive internal combustion engines. A related issue is that sufficiency varies depending on the subject and situation, meaning that one person's subjective or objective assessment of a sufficient energy supply may differ from another's under different circumstances.

7.3.2 *Energy Poverty*

If we want to live in a just world, we must decide whether we should divide energy resources as evenly as possible or simply meet everyone's fundamental needs. When the bare minimum of energy sufficiency is not achieved, energy poverty ensues. Fuel poverty is a related concept to consider.

The phrase ‘fuel poverty’ was first used in the UK more than 25 years ago to describe families whose energy bills take up more than 10% of their income (Hinson & Bolton, 2023). Since then, many countries have started using the words ‘fuel poverty’ and ‘energy poverty’ to characterise homes that, for a variety of reasons, do not have access to the energy services required to heat, cool, and power various electrical equipment. High energy prices and limited access to energy services often lead to energy poverty. For instance, inadequate infrastructure can impact a home’s ability to maintain sufficient warmth.

Energy poverty has a wide range of consequences for people’s quality of life. People who live in chilly, damp homes are more likely to get respiratory ailments. A constant cold at home has a negative impact on both physical and emotional health. When children can’t find a warm place to do their homework, it also makes it harder for them to learn. Many people feel ashamed and embarrassed when they don’t have enough money for energy. In the worst circumstances, this issue may even be fatal. Mitchell (2023) says that more than 4000 people died in England and Wales in 2022 because their homes were too cold.

People who do not have enough money for energy must make tough housing and energy use decisions. Studies have indicated that daily decisions pertain primarily to turning on the radiators or heating a hot meal, washing clothes or taking a shower, heating the entire house, or only one or two rooms. Living under such circumstances forces individuals to keep track of multiple things at once, which can make them feel stressed all the time. Studies from other countries have also found that those who are already in an unequal socioeconomic position are more likely to experience energy poverty (Mitchell, 2023).

7.3.3 Other Justice-Based Concerns Related to Energy Transition

So far, we have offered just a brief outline of the vast topic of energy transition, disregarding specific political and technological intricacies. Before proceeding, we will attempt to present a non-technical summary of the technological means of the energy transition, with a particular emphasis on the political actions required to put the means into use.

Surprisingly or not, governments around the world are currently heavily subsidising the fossil fuel industry despite the fact that renewable energy sources are often more expensive. This creates conditions in which fossil fuel businesses profit from their operations while the general public endures the environmental and social costs. To make sure that costs and benefits are shared fairly, governments should phase out fossil fuel subsidies while supporting the research and deployment of renewable energy sources. This will assist in levelling the playing field and making renewable energy sources more competitive with fossil fuels. Another key aspect is that the transition to renewable energy sources may result in increasing energy prices, which would disproportionately impact low-income households (National Conference of State Legislatures (NCSL), 2022).

Various ethical and economic considerations are driving the development and implementation of new, affordable technology to address major environmental problems around the world. The World Economic Forum identified and classified relevant technical solutions in a 2020 briefing paper. The solutions are divided into three groups, as follows:

1. Increasing energy efficiency: Recent studies show that just 33 per cent of the fundamental energy used is turned into useful energy; the remaining 67 per cent is lost due to inefficiencies in transportation, heavy industry, electricity generation, and buildings. Energy efficiency is thus among the best ways to reach an inclusive, sustainable, reasonably priced, safe energy system. Some areas in energy use have seen significant improvement (e.g. personal vehicles and aviation), whereas others (e.g. buildings) require more work.
2. Developing low-carbon energy alternatives: Well-known examples include switching from internal combustion engines to electric cars, as well as from fossil fuels to renewable energy sources such as wind and solar. However, scalable alternatives for kerosene in aviation, bunker fuels in ships, and naphtha in chemical manufacturing are still being developed.
3. Reducing unavoidable emissions: Finally, technological, economic, political, or sociological concerns may prevent the use of alternative low-carbon energy sources or the improvement of energy efficiency. In these situations, we should use carbon capture, either through carbon storage technology or natural alternatives such as reforestation.

However, no technical solution alone will solve all of the challenges related to energy transition; energy justice and the unequal distribution of economic resources must also be considered. As part of this broader justice-based solution, the benefits of the energy transition, such as reduced pollution and lower energy costs, should be distributed equitably. Otherwise, some communities are at risk of bearing a disproportionate share of the costs, such as the environmental impacts of renewable energy technologies.

Specifically, switching to renewable energy sources may require significant infrastructure investments, which could be difficult to finance in low-income areas (Lennon et al., 2019). Furthermore, renewable energy solutions like solar panels and energy-efficient appliances must be affordable and easily accessible. If these technologies are not accessible to all income groups, there is a risk of creating an energy divide where wealthier individuals benefit more from the transition.

Among other justice-based issues concerning energy transition, the shift from fossil fuels to renewable energy sources may result in job displacement in coal mining and oil extraction, among other industries, which might significantly impact workers' livelihoods (Simon, 2020). Concerns about social justice arise when workers in these businesses, who are generally from underprivileged communities, are not appropriately supported with retraining and transition schemes.

A strong concept of social justice requires that all communities, especially marginalised groups (be they women, disabled people, or ethnic minorities), participate in energy transition decision-making processes. In these processes, a lack of inclusion can lead to policies that do not address the unique needs and concerns of different people. Environmental concerns linked with energy production and extraction disproportionately affect specific populations, most of whom are low-income (Blanchet & Martínez-Toledano, 2023). Transitioning to cleaner energy sources should not perpetuate or exacerbate existing economic and environmental injustices.

Indigenous communities may face unique challenges in the energy transition, as their land and cultural practices may be directly affected by renewable energy projects or other forms of energy development. Ensuring respect for cultural rights and engaging in meaningful consultation with these communities is essential.

The benefits of the energy transition should not be concentrated in certain regions while others are left behind. The Global South countries may face challenges in adopting clean energy technologies due to financial constraints, and efforts should be made to support them in their transition. A holistic and inclusive approach to policymaking is necessary to address these social justice concerns, with a focus on ensuring a fair distribution of the benefits and burdens of the energy transition among all citizens. Therefore, it is crucial to ensure that the public has a say in how the transition is planned and implemented. This includes making sure that the public is involved, that the community has a say in planned projects, and that the decision-making process is transparent.

Furthermore, the energy transition cannot occur in just one country or region. It is a global challenge requiring worldwide cooperation. This entails addressing climate change on a global basis and ensuring an equal transition for all countries. The energy transition should not come at the expense of environmental justice. This includes ensuring that the site of renewable energy projects does not harm the environment or local residents. It also means addressing the legacy of pollution and environmental damage caused by the fossil fuel industry.

Governments, civil society organisations, and the private sector all contribute to a socially just and equitable transition to renewable energy (Hafner & Tagliapietra, 2020). It should be underlined that not only do financial resources differ among countries, but so do cultural and political ideas on social justice, complicating efforts to address inequities in the global energy transition. The debate over the intricate connection between ecological sustainability and socially equitable distribution illustrates this.

7.3.4 Ethical Frameworks Related to Energy Transition

The energy transition generates numerous ethical concerns and a need for policy recommendations. As stated, the use of fossil fuels contributes to climate change, which poses a significant threat to biodiversity. This is what drives the energy transition. As a result, the energy transition is meant to protect the climate and nature from damage caused by people. Several aspects of this endeavour are worthy of consideration from an ethical point of view. Let us give a few examples.

A key principle of deontological ethics is the idea that some actions are morally obligatory, regardless of the consequences. In the case of the energy transition, this framework would argue that governments, companies, and individuals have a moral duty to reduce greenhouse gas (GHG) emissions because it is intrinsically wrong to harm the environment and future generations, irrespective of the economic costs or benefits of such actions.

From the perspective of utilitarian or consequentialist ethics, the benefits of the energy transition are what makes it important. According to Jeremy Bentham (1970), the criterion for the rightness of an action is the benefit it produces. Therefore, the utility of an action's effects, rather than the decency of its aims, should ultimately define its goodness. Bentham saw human happiness as the ultimate utility. He came to the conclusion that the 'greatest happiness for the greatest number' is what we ought to strive for (Bentham, 1977: 393, 440n).

Bentham's utility principle could also be used to frame the energy transition: Produce as much environmentally friendly energy as possible for as many people as possible. While this may sound attractive, there are problems with the utility principle as applied to the energy transition. The largest amount of environmentally friendly energy for the largest number of people may be a counterproductive goal. The resulting increase in consumption may be beyond the ecological carrying capacity of the planet while potentially meeting the lower limit of Raworth's (2017) doughnut economy: The social minimum of meeting basic human needs.

The third classical model is virtue ethics, which is equally relevant to the energy transition. Virtue ethics focuses on developing character traits necessary for a life with ethical integrity. Thus, virtue ethics highlights the moral agent's character as a critical component of ethics rather than duties or outcomes. Virtue ethics holds that in order for people to develop positive character traits like moderation and fairness, they need to be raised and educated. In today's world, when nature's carrying capacity is under threat and the distribution of wellbeing among various individuals is jeopardised, moderation and fairness are especially important. Let's consider the CEO of a large energy company such as Exxon Oil, Chevron, or PetroChina. If the CEO adopts a virtue ethics perspective when making decisions about the energy transition, they would not solely prioritise profits or adhere to legally mandated standards. Rather, they would actively foster a sense of responsibility for a variety of stakeholders, such as the environment, future generations, local communities, and employees.

7.4 Navigating Social Justice in Renewable Energy Systems: The Role of the Supply Chain in a Just Transition

Renewable energy technologies (RETs) are pivotal in driving the transition towards clean and sustainable energy systems. Recent advancements in technology, improved efficiencies, reduced costs, and the potential for small-scale distributed energy generation have brought RETs to the forefront of energy transition (IEA 2023b; IRENA, 2023). However, despite offering environmental, economic, and social benefits, widespread adoption of RETs faces significant challenges (Shakeel et al., 2017).

The manufacturing and integration of RETs require critical raw materials and minerals, without which their development would not be possible. These include silicon for solar panels, rare-earth minerals for wind turbines, and lithium, cobalt, and nickel for batteries and other RETs (Giurco et al., 2019; Sovacool et al., 2020). Ensuring an uninterrupted extraction and supply of these essential resources is essential for a continued development and application of RETs. Many of these minerals are located in remote areas and geographically concentrated in specific regions of the world, often far from where manufacturing takes place, which is a critical aspect of mineral extraction (Sovacool et al., 2020). This necessitates a robust and functional supply chain to transport these minerals to manufacturing locations, making the supply chain an essential element for RETs availability and affordability.

The RET supply chain encompasses several stages: Raw material extraction, manufacturing, installation and deployment, operation and maintenance, and end-of-life handling. Mining operations perform the extraction, obtaining necessary resources that undergo refinement and processing for technology development. Manufacturing is the process of producing components and assembling technologies. Installation and maintenance focus on the application of RETs and ensuring their ongoing functionality. The end-of-life handling process entails managing the disposal or recycling of components and systems that have reached their useful life. Each stage involves different actors and stakeholders, including policymakers, regulators, companies, local communities, and nongovernmental organisations, and presents unique challenges that influence aspects of justice.

Transitioning from a conventional to a renewables-based energy system necessitates addressing justice issues at each step to guarantee that the transition is just and equitable for all actors and stakeholders across the entire value chain. They must integrate justice considerations from the outset and not only address them at the final stage of applying renewable energy technologies. Neglecting justice implications during the extraction process means that efforts to address challenges and implement remedies will be inherently flawed, allowing early-stage injustices to persist throughout the life span of RETs.

Researchers have devised numerous frameworks and methodologies to investigate the extent to which justice aspects are considered in energy settings. Sovacool and Dworkin (2015) introduced a principle-based approach to energy justice, including availability, affordability, due process, transparency and accountability, sustainability, intergenerational equity, and responsibility. Subsequently, Sovacool et al. (2017) added resistance and intersectionality as additional justice principles. Availability ensures individuals and communities have reliable access to energy, while affordability emphasises fair pricing, especially for lower-income and vulnerable populations. Due process demands fair and equitable procedures, and transparency and accountability guarantee open systems that hold actors accountable for their actions. Sustainability encompasses environmental factors, while intergenerational equity safeguards the needs of future generations. Responsibility acknowledges the impact on individuals and communities, requiring actors to take responsibility for their actions. Resistance recognises individuals' and communities' rights to oppose projects with negative consequences, while intersectionality considers the overlap and intersections of various justice aspects. These principles collectively aim to foster an equitable and sustainable energy transition by acknowledging different impacts and considerations relevant to the decisions and actions (Tolonen et al., 2023).

Similarly, another influential framework in the study of energy justice is the tenet-based approach to justice, including distributive justice, procedural justice, and recognition justice (Jenkins et al., 2021). Scholars have widely adopted this framework, often referred to as the triumvirate of energy justice, to delve deeper into energy justice issues (McCauley et al., 2013). Distributive justice guarantees a fair distribution of benefits and costs from energy projects among the actors and stakeholders (McCauley et al., 2019). Procedural justice emphasises inclusive and fair decision-making processes, particularly considering vulnerable and marginalised communities. It also underscores the importance of transparency in sharing information and the adoption of fair practices and procedures (Jenkins et al., 2016; Milchram et al., 2020). Recognition justice asserts that we should consider the needs and concerns of all actors and stakeholders and not ignore or under-represent any part of society (Schlosberg, 2004). It also emphasises the importance of addressing historical and existing injustices. Siciliano et al. (2018) and McCauley and Heffron (2018) build on the framework and advocate for an additional dimension focused on restoration activities to rectify harm caused by energy activities. Restorative justice acknowledges the need to implement measures that prevent harm to society, the environment, or the ecosystem due to energy projects. This underscores the importance of a proactive approach to establishing restorative measures that mitigate environmental damage and address challenges resulting from the activities.

The following explores how justice issues manifest in the exploration phase of the renewable energy supply chain (Table 7.1).

7.4.1 Distributive Justice

Distributive justice is critical in the mineral extraction process for RET development. Ensuring a fair distribution of benefits and costs arising from extraction activities establishes the foundation for fairness and equity throughout the process. Many of

Table 7.1 Justice issues in renewable energy extraction processes

Justice consideration	Manifestation in renewable energy extraction processes
Distributive justice Equitable and fair distribution of benefits and costs among all actors and stakeholders	– Inequitable distribution of benefits between companies and stakeholders – Fairness, openness, and transparency about benefits, with financial incentives passed to local communities and stakeholders – Negative environmental externalities, such as contamination of water, land degradation, and pollution – Loss of local biodiversity affecting livelihoods and economic activities of the residents – Inadequate compensation for damages and harms suffered – Limited employment and business opportunities for locals – Lack of development or underfinancing of local infrastructure support – Minimal or no contribution towards the local society from companies – Inadequate investment in community services, e.g. education and healthcare
Procedural justice Access to and participation in decision-making processes that guide activities with an emphasis on fairness and transparency	– Lack of transparency in permitting processes for extraction activities – Insufficient community engagement and consultation efforts – Bias in the engagement process, limiting participation to first movers who are typically more knowledgeable and resourceful, leading to flawed procedural outcomes – Exclusion of marginalised groups from decision-making processes – Flawed application of regulations and policies – Manipulation of processes, laws, and practices to gain favourable outcomes – Inadequate practices for community feedback and a desire to overhaul processes based on recommendations – Limited public consultations or discussion before actual project implementation – Failure to involve local leaders and community representatives in planning processes

(continued)

Table 7.1 (continued)

Justice consideration	Manifestation in renewable energy extraction processes
Recognition justice Acknowledgement, participation, and appreciation of stakeholders and actors, especially vulnerable, marginalised communities	– Lack of recognition for vulnerable, and marginalised communities – Ignorance of local customs and traditions – Overlooking the cultural significance of local lands and indigenous communities – Lack of recognition for locals' rights and failure to honour responsibilities towards them – Stereotyping or misrepresentation of local communities in project narratives – Neglect of traditional knowledge and their spiritual connection to local customs, traditions, and land – Limited support for local cultural initiatives – Disregard for historical injustices faced by locals
Restorative justice Repairing harm and reinstating the state to minimise damage caused by the establishment and operationalisation of the energy projects	– Unclear or weak grievance mechanisms for local residents – Limited reparations for communities impacted by operations – Commitment to environmental restoration and repairing harm caused by activities – Community led monitoring and evaluation of environmental impacts – Educational programmes to build trust and understanding between stakeholders – Capacity building to prevent accidents, with clear compensation mechanisms and legal actions for responsible parties – Neglect of rehabilitation and improvement for local surroundings – Failure to acknowledge past issues and undertake preventive actions to avoid future accidents – Active engagement of affected communities in designing restorative actions – Regular assessment of social impacts to inform ongoing project management

these essential minerals are found in remote areas far from urban centres, where people live in small, rural settlements. Large national or multinational corporations, prioritising profit-making and lacking local dynamics, often carry out complex extraction processes, requiring sophisticated technology and equipment.

As these companies aim to expand extraction and generate revenues, a significant portion of the profits often translates into financial gains for the companies, with very little returned to local communities and society. This disparity often leads to contention, as communities view these vital, depletable resources as potential means to improve their wellbeing and prosperity. While companies reap substantial profits, the extraction process frequently degrades the environment and disrupts local ecosystems, making daily life difficult for communities.

For instance, lithium extraction in the Atacama Desert in Chile has raised several distributive justice concerns (Jerez et al., 2021; Liu & Agusdinata, 2020). The global increase in demand for electric batteries has driven the need for lithium extraction, making it a major economic activity in the region (Jerez et al., 2021). However, the lack of fair distribution of benefits to local communities, coupled with adverse environmental impacts such as depletion of local water resources and harm to agriculture and traditional livelihoods, has sparked conflict and legal disputes (Babidge, 2019; Jerez et al., 2021). Similarly, cobalt mining in the Democratic Republic of Congo (DRC) illustrates distributive justice issues. Local communities criticise the extraction process, noting that community members, miners, and the local ecosystem suffers health issues and environmental degradation without receiving adequate economic returns. Artisanal miners, including children, endure hazardous conditions without receiving their fair share, while the majority of the profits accrue to the companies conducting operations (Katz-Lavigne, 2019; Sovacool, 2019).

These examples of justice are not just limited to the global south countries or developing economies alone, rather, members and actors from the developed countries also suffer from similar injustices. The Voisey's bay Nickle mine in Canada highlights distributive justice concerns resulting from the mineral extraction. The mine, located on indigenous lands, had significant impacts on the local communities. The concerns of environmental degradation related to water and wildlife, negative impacts on traditional ways of living, and harmful long-term impacts on the surroundings have caused conflict among the locals and the parties involved in extraction processes (Gibson, 2006). Similar concerns were raised at Talvivaara Nickle and Cobalt mine in Finland. The mine is home to nickel and cobalt used in various RETs and equipment. The leak of toxic metals into local waterways led to severe pollution in surrounding lakes and rivers. The contamination resulted in loss of fish stock, contaminated water supplies, and significant health issues. Residents registered reservations that they were inadequately compensated for the damage, have greatly suffered from the negative outcomes, while external parties and investors benefited from the economic yields resulting from the mining activities (Sairinen et al., 2017; Tiainen et al., 2014).

These disparities highlight the need for measures to ensure that the local community benefit from mineral extraction related activities through economic gains and improved living conditions. All key stakeholders should be involved in developing strategies and plans to improve the local ecosystem, broaden knowledge bases, and offer community members better job opportunities beyond labour-intensive tasks. When approached with a focus on equity, projects can contribute to the local community and wellbeing. The Reglan mine, Quebec, highlights positive initiatives undertaken to ensure the distribution of benefits to the local communities. An agreement was signed between the extraction company and a local community owned economic development company, Makivvik Corporation, along with five Inuit partner corporations. The agreement emphasised environmental stewardship, local empowerment, and priority to local business corporations for procurement. The agreement also ensured profit-sharing arrangements and a commitment to provide 4.5 per cent of operating profits once the mine has recouped initial capital investment. The accumulated funds were then distributed to local communities (Reglan ICMM, 2022; Mine, 2021; SIFEE, 2023). Measures of such nature improve the sense of equity and fairness in the distribution of benefits.

7.4.2 Procedural Justice

The principles of inclusivity and fairness should guide decision-making processes concerning the extraction of mineral resources, as emphasised by procedural justice. A participatory and inclusive approach ensures that diverse perspectives are considered, thereby increasing the likelihood that the distribution of benefits and costs adheres to the principles of fairness. Decision-making bodies often include individuals and groups with greater knowledge, resourcefulness, and interest in the issues at hand (Gillard et al., 2017; Milchram et al., 2018). However, these decision-making bodies often overlook the concerns of vulnerable and marginalised communities due to their lack of representation. Bias in the process often results in outcomes that disproportionately affect vulnerable and marginalised groups.

For instance, mining in Inner Mongolia, China, illustrates significant procedural injustices. The decision-making processes that govern the mining projects have excluded local communities, including herders, from meaningful participation. Government officials and large corporations dominated these processes, undermining local population concerns regarding land use, protection, and community impacts. Such procedural injustices have led to the displacement of herders and the disruption of their traditional livelihood, fostering community resentment and distrust throughout the entire process. (Denyer, 2015; Woodworth, 2017).

This pattern of exclusion is also noticeable in other cases, such as the Philippines, where nickel mining highlights cases where local communities were sidelined in consultations and decision-making processes concerning mining activities. Mining interests often manipulated environmental impact assessments and community consultations, leading to environmental hazards like water pollution, deforestation, and detrimental impacts on the local ecosystem, which significantly affect communities' livelihoods. The neglect of procedural justice issues undermines genuine participatory engagement of local communities and their ability to have their voices and concerns heard, thereby diminishing fairness and transparency (Holden et al., 2011; Nem Singh & Camba, 2016). Rare earth elements mining in Greenland also showcased several procedural justice issues. Limited public participation despite the project's significant environmental impacts, lack of transparency, worries concerning social impact assessment processes, and political interferences to proceed with the project despite serious environmental concerns highlight the case where process and procedures were influenced to gain favourable decisions (Bjørst, 2016; Henriques & Böhm, 2022; Reuters, 2021).

Similarly, lithium extraction at Argentina's Salar de Olaroz project highlights concerns regarding procedural issues. Local communities repeatedly voiced concerns about deteriorating environmental conditions and negative impacts on agriculture and cultural practices. Despite their concerns, the decision-making processes did not fully integrate community input, raising concerns that the decisions favoured mining operations over the protection of local resources (Díaz Paz et al., 2023; Dorn, 2021; Escosteguy et al., 2022). This raises issues of fairness and transparency in the process, leading to disputes.

On the contrary, companies' efforts to strengthen procedural aspects can smoothen development and ensure that the community benefits from extraction activities. For instance, efforts to enhance community engagement and increase local actors' participation at the Salobo copper mine in Brazil helped relieve the conflict. Vale, the mining company, initiated a close collaboration with indigenous groups to the mine's impacts on the local national forest, which is home to various Indigenous tribes and is an important element of the local ecosystem can be mitigated. A multi-stakeholder committee was also established to oversee the environmental and social impacts of the project. Additionally, measures were designed to increase transparency in decision-making processes, enabling affected communities to register their concerns and provide input on mining operations. These activities played a part in engaging local communities and ensuring that their cultural and environmental concerns were considered (Vale, 2021, 2022). However, a lot more is required to strengthen engagement and address community concerns (Saes et al., 2021).

7.4.3 Recognition Justice

Recognising vulnerable and marginalised communities, as well as the impact of mining activities on their wellbeing, is a critical aspect of recognition justice. This is essential because, intentionally or unintentionally, ignoring certain groups of individuals, communities, and members of society means that their voices, concerns, and valuable contributions remain unheard. This sets the stage for future injustices and challenges to transparency and fairness.

For example, in the case of bauxite extraction in Guinea, a major producer of minerals used in several RETs, significant recognition justice issues arise. Local communities, reliant on agriculture and fishing, see their livelihood threatened by the environmental impact of the mining activities. Water contamination and degradation of the farmland severely impact their wellbeing (Camara et al., 2021). The lack of recognition of these local people and communities has contributed to the decision-making processes and activities that have led to serious consequences for the well-being of vulnerable and marginalised communities (Dibattista et al., 2023). This has fuelled conflicts and resentment towards mining activities (Bolay, 2014; Knierzinger, 2018). A collaborative approach involving all key actors and stakeholders, establishing pathways for sustainable living and protection of the environment, is essential to address these challenges.

Similarly, Mozambique's graphite mining operations present several issues related to recognition and justice. Mining activities in Cabo Delgado province, one of the largest producers of graphite used in RETs, have displaced communities as their local lands and resources were taken over by companies for mining operations without proper consent and compensation mechanisms (Kirshner & Power, 2015; Wiegink, 2020). This has resulted in the loss of assets and sources of income for affected groups. The lack of attention to local issues has fostered discomfort and diminished trust in the actors and stakeholders, negatively impacting community confidence. Therefore, key stakeholders must acknowledge the concerns of vulnerable and marginalised communities and implement alternative strategies to sustain their livelihoods (Hilson et al., 2021). The mining operations should provide adequate compensation for their lands and vital resources.

For example, the Oyu Tolgoi mine in Magnolia illustrates measures undertaken by the development companies to address local communities' concerns about inadequate consultation and representation of the interests of local herders and the general public. The companies' operations faced resistance as the community felt they were not adequately recognised as stakeholders, and their concerns were unheard. This fuelled mistrust between the company, the government and the local communities regarding the operations. To address these challenges, efforts were made to ensure the inclusion of nomadic herders' concerns in the decision-making, recognising the importance of indigenous and traditional knowledge and cultural identity, as well as involving herders in consultation and negotiation about local resource management. This offered marginalised communities a platform to express their concerns and propose solutions (Tinto, 2024). However, many called for more extensive measures and an in depth understanding of the local dynamics to improve the justice considerations (Oyu Tolgoi Watch, 2012). The recognition of key actors and stakeholders in the decision-making processes and the impacts in terms of consequences can act as a foundation based on which a project's fairness can be determined.

7.4.4 Restorative Justice

Because mining operations have significant environmental and social impacts, restorative justice is critical in mineral extraction. Extracting minerals often requires extensive use of water resources and can lead to adverse environmental consequences for the land, surrounding communities, and biodiversity. Several restorative justice issues have emerged in the case of nickel mining activities in Norilsk, Russia. Decades of nickel mining have resulted in extensive environmental pollution and contamination of air and water sources (CBEES, 2023). The heavy operations have also caused noise pollution and loss of property, significantly affecting the livelihood, health, and wellbeing of local indigenous communities. Addressing restorative justice issues is central to ensuring that companies and stakeholders commit to improving environmental conditions and restoring affected areas to their original state, both during and after mining (Yakovlev et al., 2008; Bellona, 2023; Norisk Nickel, 2023). This includes efforts such as water purification, reforestation, and biodiversity restoration. Furthermore, it is crucial for companies to consider and respect the local culture and values of indigenous communities. The Nenets and Sámi, indigenous communities of Norilsk, have strong ancestral ties to their lands. Their orientation and relationship with the land often differ significantly from individuals coming from urban settings. Therefore, all stakeholders must carefully consider these differences and ensure that restorative actions address these cultural factors to conduct restoration in a manner that meets the needs of local communities.

Similarly, tin mining in the Bangka-Belitung region of Indonesia has raised serious restorative justice challenges. Significant environmental damage, including coastal erosion, deforestation, loss of natural resources, and damage to coral reefs, has been a major concern. Furthermore, the loss of income from local finishing activities and agriculture has been a significant factor. To mitigate these effects, local NGOs, governmental agencies, and local stakeholders have collaborated on restoration efforts (Ibrahim et al., 2018; Rosyida et al., 2018; Darwance et al., 2023). These initiatives focus on repairing coastal damage, promoting sustainable fishing practices, and training local communities in sustainable fishing and agriculture. Such efforts not only help address past damages but also build resilience and foster a sustainable future for affected communities.

The Talvivaara mine restoration in Finland further showcases how restorative measures can minimise the effects of extraction activities. The environmental challenges in the form of water pollution and metal emissions, soil contamination, and disruption of the local ecosystem raised restorative justice issues. The mine restoration activities focused on wastewater management and treatment through the construction of own central water treatment plant and the creation of an area for passive water purification. Actions were also taken to ensure biodiversity conservation through the development of biodiversity offset areas, the introduction of native plant species, and the creation of artificial habitats for displaced wildlife (Förvaltningsdomstolen; STUK, 2024; TEM, 2017). These measures focused on restoring wildlife, natural ecosystem, water quality, and mitigating the environmental damage caused by the extractive operations.

In addition to corrective measures, adopting a proactive approach can significantly strengthen restorative justice. Rather than waiting for extractive activities to conclude before initiating the restoration process, it is essential to plan and implement restorative actions from the outset. This includes taking measures to prevent environmental and livelihood damage during the operations, establishing clear corrective actions, identifying responsible authorities in the event of accidents, and providing clear guidance on correction plans and remedies for the affected communities.

7.5 Policy Recommendations for the Energy Transition

The examination of current publications has helped us to classify the social issues related to the energy transition into five main groups: Rising energy prices, energy poverty, job displacement, and unemployment due to the energy transition, unequal benefit distribution, and gender and ethnic injustices caused by the transition. This distinction reflects a risk- and problem-orientated approach to energy justice. However, when discussing justice in the energy transition, it is also essential to consider the various aspects of justice stated earlier, including distributive justice, procedural justice, recognition justice, and restorative justice. In response to various risks and problems, as well as different dimensions of justice, we can now propose the development and implementation of the following policy recommendations, which address critical challenges in the energy transition from various social justice standpoints. We frame the following recommendations as deontological orders and duties, but we can also justify them in terms of consequential ethics based on the anticipated environmental and societal benefits of their implementation. In each recommendation, we additionally include, in square brackets, the dimension of justice that the recommendation specifically embodies. For each recommendation, we provide a real-life example of a successful policy:

1. Governments across the board should phase out subsidies for fossil fuels while simultaneously promoting the development and use of renewable energy sources throughout the supply chain. [Procedural justice]

   Example: Canada, a major fossil fuel producer, has committed to phasing out inefficient fossil fuel subsidies by 2025 as part of its commitments under the G7 and G20 frameworks. In 2022, the Government of Canada started scaling back subsidies for oil and gas companies, especially those that support exploration and extraction activities (Government of Canada, 2023).
2. All income groups must have access to clean energy technologies like solar panels and energy-efficient appliances. In this, the supply chain is critical and needs to be carefully considered. [Distributive justice]

   Example: In the United States, the Low-Income Home Energy Assistance (LIHEAP) Weatherisation Assistance Programme provides funding to help low-income households improve their home energy efficiency, reduce energy bills, and increase comfort. This includes installing energy-efficient appliances, insulating homes, and upgrading heating and cooling systems (Office of Community Services, 2024).
3. Workers in industries like coal mining and oil extraction must be adequately supported with retraining and transition programmes. [Restorative justice]

   Example: Spain has committed to shutting down all coal mines and coal-fired power plants by 2030. To mitigate the impacts on workers and communities dependent on coal, the Government of Spain has developed a comprehensive Just Transition Strategy (World Resources Institute, 2021).

4. Social justice requires that all communities have a say in the decision-making processes related to the energy transition. Ensuring respect for cultural rights and engaging in meaningful consultation with indigenous communities is essential. [Procedural justice and recognition justice]

   Example: In Canada, several provinces, including Ontario and British Columbia, have programmes that provide financial support and technical assistance to indigenous communities to develop and own renewable energy projects. For example, Ontario's Aboriginal Renewable Energy Fund helped indigenous communities participate in the development of solar and wind projects (Mongabay, 2024).
5. The benefits of the energy transition, such as reduced pollution and lower energy costs, should be distributed equitably. Transitioning to cleaner energy sources should not perpetuate or exacerbate existing environmental injustices. [Distributive justice]

   Example: In South Africa, the Integrated National Electrification Programme (INEP) initiative aims to provide universal access to electricity, with a focus on renewable energy. The Government of South Africa has been deploying solar home systems in rural areas that are not connected to the national grid, ensuring that low-income households have access to reliable and clean energy (Energypedia, 2018).
6. It is important to assist the Global South countries with their energy transition. [Distributive justice and recognition justice]

   Example: The Green Climate Fund (GCF) is the world's largest dedicated climate finance fund, created under the United Nations Framework Convention on Climate Change (UNFCCC) to support developing countries in mitigating and adapting to climate change, including energy transitions (United Nations Climate Change, 2024).
7. Governments, civil society, and businesses should all be involved in ensuring a socially just and equitable transition to renewable energy. [Procedural justice and recognition justice]

   Example: The Africa Clean Energy Corridor, led by the International Renewable Energy Agency (IRENA), involves partnerships between governments, civil society, and private businesses to scale up renewable energy development in East and Southern Africa. The initiative focuses on ensuring affordable and reliable energy access while promoting job creation and social inclusion (IRENA, 2022).

These recommendations extensively consider various aspects of justice. As a result, justice in the energy transition includes the realisation of the right to speak, the adoption of inclusive procedures, the rectification of historical injustices, and the fair distribution of costs and gains. These extensive policies require collaboration between different stakeholders. Furthermore, tailoring the recommendations should take various national and international settings into account. In summary, we must understand the recommendations inclusively and contextualise them.

7.6 Conclusion

We must transition to a low-carbon economy in order to mitigate the climate crisis, but we must also remember that this transition is not just about technology and economics; it is also about social change. We have seen that addressing a number of key issues is necessary to make the energy transition socially just. These considerations have been offered to assist academic researchers and practitioners, including policy planners and community developers, in integrating communality activities and the energy transition.

In terms of social justice, the biggest challenge in the energy transition is fairly distributing benefits and expenses to various groups of people across the renewable energy supply chain. The negative environmental impacts of energy production and consumption are often borne by marginalised communities, such as indigenous peoples and people of colour. Therefore, it is crucial to ensure that the public, understood broadly, has a say in how the transition is planned and implemented. This means ensuring that there is public participation in decision-making processes, that communities are consulted about proposed energy transition projects, and that there is transparency in the decision-making process.

As countries transition to cleaner energy sources, there is a risk that vulnerable populations may be left behind. The transition to renewable energy sources may result in higher energy prices, disproportionately affecting low-income households. Affordability and access to clean energy technologies are crucial to ensure that all members of society can benefit from the transition.

The transition to renewable energy sources typically requires major infrastructure investments across the supply chain, which may be difficult to finance in low-income areas. Furthermore, the transition to renewables may result in job losses in the fossil fuel industry, having a severe impact on workers' livelihoods. This necessitates providing targeted support to help these groups access clean energy and ensure that they are not disproportionately affected by the transition.

The energy transition should not come at the expense of environmental justice. This means ensuring that the locations of renewable energy projects have no negative impact on the environment or local residents. It also means addressing the legacy of pollution and environmental damage caused by the fossil fuel industry.

Dealing with these challenges calls for a comprehensive approach considering the interests and concerns of all stakeholders, including low-income households, fossil fuel industry workers, and marginalised communities. Ensuring a socially just and equitable transition to renewable energy is a shared responsibility of governments, non-governmental organisations, and the corporate sector.

The energy transition cannot be a purely national or regional endeavour. It is a global challenge that requires international cooperation. This means tackling climate change on a global basis and ensuring that the transition is fair to all countries. The energy transition could benefit everyone by addressing these critical issues.

References

Babidge, S. (2019). Sustaining ignorance: The uncertainties of groundwater and its extraction in the Salar de Atacama, northern Chile. *The Journal of the Royal Anthropological Institute, 25*(1), 83–102. https://doi.org/10.1111/1467-9655.12965

Balachandra, P., Nathan, H. S. K., & Reddy, B. S. (2010). Commercialization of sustainable energy technologies. *Renewable Energy, 35*(8), 1842–1851. https://doi.org/10.1016/j.renene.2009.12.020

Bantham, J. (1977). *A comment on the commentaries and a fragment on government*. Oxford University Press.

Bellona. (2023). Industrial pollution in the Russian Arctic is an environmental nightmare: A list of the dirtiest places. https://bellona.org/news/industrial-pollution/2023-11-industrial-pollution-in-the-russian-arctic-is-an-environmental-nightmare-a-list-of-the-dirtiest-companies

Bentham, J. (1970). *An introduction to the principles of morals and legislation*. Methuen.

Bjørst, L. R. (2016). Saving or destroying the local community? Conflicting spatial storylines in the Greenlandic debate on uranium. *The Extractive Industries and Society, 3*(1). https://doi.org/10.1016/j.exis.2015.11.006

Blanchet, T., & Martínez-Toledano, C. (2023). Wealth inequality dynamics in Europe and the United States: Understanding the determinants. *Journal of Monetary Economics, 133*, 25–43. https://doi.org/10.1016/j.jmoneco.2022.11.010

Bolay, M. (2014). When miners become "foreigners": Competing categorizations within gold mining spaces in Guinea. *Resources Policy, 41*(1), 117–127. https://doi.org/10.1016/j.resourpol.2014.02.001

Bonnedahl, K. J., & Heikkurinen, P. (2019). *Strongly sustainable societies: Organising human activities on a hot and full earth*. Routledge.

Camara, I., Deyi, J., Barry, O., & Caille, F. (2021). Bauxite mining conflicts in Guinea: Causes identification, analysis, and countermeasures. *International Journal of Mineral Processing and Extractive Metallurgy, 6*(3), 53–66. https://doi.org/10.11648/j.ijmpem.20210603.13

CBEES. (2023). Ecological concerns in transition—A comparative study on responses to waste and environmental destruction in the region.

Darwance, H. D., Sari, R., Anwar, M.S., & Satrio, N. (2023). Tin mining in Bangka Belitung Islands and its impact on the reputation of geographical indication: A policymakers perspective. In *IOP conference series: Earth and environmental science* (pp. 0102011).

Denyer, S. (2015). In China's Inner Mongolia, mining spells misery for traditional herders. In Washington Post. https://www.washingtonpost.com/world/asia_pacific/in-chinas-inner-mongolia-mining-spells-misery-for-traditional-herders/2015/04/07/16b3a252-d643-11e4-bf0b-f648b95a6488_story.html

Díaz Paz, W. F., Escosteguy, M., Seghezzo, L., Hufty, M., Kruse, E., & Iribarnegaray, M. A. (2023). Lithium mining, water resources, and socio-economic issues in northern Argentina: We are not all in the same boat. *Resource Policy, 81*.https://doi.org/10.1016/j.resourpol.2022.103288

Dibattista, I., Camara, A. R., Molderez, I., Benassai, E. M., & Palozza, F. (2023). Socio-environmental impact of mining activities in Guinea: The case of bauxite extraction in the region of Boké. *Journal of Cleaner Production, 387*. https://doi.org/10.1016/j.jclepro.2022.135720

Dorn, F. M. (2021). Changing territorialities in the Argentine Andes: Lithium mining at salar de olaroz-cauchari and salinas grandes. *Erde 152*(1). https://doi.org/10.12854/erde-2021-515

Droege, P., & Radz i A, Carlisle N, Lechtenbohmer S,. (2010). *100% Renewable energy and beyond for cities*. Hamburg.

Energypedia. (2018). NAE case study: South Africa, integrated national electrification. https://energypedia.info/wiki/NAE_Case_Study:_South_Africa,_Integrated_National_Electrification

Escosteguy, M., Diaz Paz, W. F., Iribarnegaray, M. A., Clavijo, A., Ortega Insaurralde, C., Stern, H., Venencia, C. D., Brannstrom, C., Hufty, M., & Seghezzo, L. (2022). Will electro-mobility encourage injustices? The case of lithium production in the Argentine Puna. In *Energy democracies for sustainable futures* (pp. 225–232).

European Commission. (2023). State of the Energy Union Report. https://eur-lex.europa.eu/legal-content/EN/TXT/?uri=COM%3A2023%3A650%3AFIN&qid=1698237100377
Förvaltningsdomstolen, H KHI. (2017) 76.
García-García, P., Carpintero, Ó., & Buendía, L. (2020). Just energy transitions to low carbon economies: A review of the concept and its effects on labour and income. *Energy Research and Social Science, 70*, 101664. https://doi.org/10.1016/j.erss.2020.101664
Gibson, R. B. (2006). Sustainability assessment and conflict resolution: Reaching agreement to proceed with the Voisey's Bay nickel mine. *Journal of Cleaner Production, 14*(3–4). https://doi.org/10.1016/j.jclepro.2004.07.007
Gillard, R., Snell, C., & Bevan, M. (2017). Advancing an energy justice perspective of fuel poverty: Household vulnerability and domestic retrofit policy in the United Kingdom. *Energy Research and Social Science, 29*, 53–61. https://doi.org/10.1016/j.erss.2017.05.012
Giurco, D., Dominish, E., Florin, N., Watari, T., & McLellan, B. (2019). Requirements for minerals and metals for 100% renewable scenarios. In S. Teske (Ed.), *Achieving the Paris climate agreement goals* (pp. 437–457). Springer.
Hafner, M., & Tagliapietra, S. (2020). The global energy transition: A review of the existing literature. *The geopolitics of the global energy transition* (pp. 1–24). Springer.
Hartmann, E. (2024). Sufficiency as relations of enoughness. *Sustainable Deviation, 32*. https://doi.org/10.1002/sd.3090
Henriques, I., & Böhm, S. (2022). The perils of ecologically unequal exchange: Contesting rare-earth mining in Greenland. *Journal Cleaner Production, 349*.https://doi.org/10.1016/j.jclepro.2022.131378
Hilson, G., Mondlane, S., Hilson, A., Arnall, A., & Laing, T. (2021). Formalizing artisanal and small-scale mining in Mozambique: Concerns, priorities and challenges. *Resourcing Policy, 71*.https://doi.org/10.1016/j.resourpol.2021.102001
Hinson, S., & Bolton, P. (2023). Fuel poverty. House of Commons. https://researchbriefings.files.parliament.uk/documents/CBP-8730/CBP-8730.pdf
Holden, W., Nadeau, K., & Jacobson, R. D. (2011). Exemplifying accumulation by dispossession: Mining and indigenous peoples in the Philippines. *Geografiska Annaler Series B-human geography, 93*(2). https://doi.org/10.1111/j.1468-0467.2011.00366.x
Hrnčić, B., Pfeifer, A., Jurić, F., Duić, N., Ivanović, V., & Vušanović, I. (2021). Different investment dynamics in energy transition towards a 100% renewable energy system. *Energy, 237*, 121526. https://doi.org/10.1016/j.energy.2021.121526
Ibrahim, H. D., & Wahyudin, N. (2018). From charm to sorrow: The dark portrait of tin mining in Bangka Belitung, Indonesia. *People International Journal of Social Sciences, 4*(1), 360–382. https:// grdspublishing.org/index.php/people/article/view/966
ICMM. (2022). Raglan mine—Accessing royalties and profit-sharing mechanisms.
IEA. (2023a). Global EV Outlook 2023—Analysis—IEA. https://www.iea.org/reports/global-ev-outlook-2023
IEA. (2023b). World Energy Outlook 2023. https://iea.blob.core.windows.net/assets/86ede39e-4436-42d7-ba2a-edf61467e070/WorldEnergyOutlook2023.pdf
ILO. (2015). Guidelines for a just transition towards environmentally sustainable economies and societies for all. https://www.ilo.org/publications/guidelines-just-transition-towards-environmentally-sustainable-economies
IRENA. (2022). Africa clean energy corridor. https://www.irena.org/Energy-Transition/Country-engagement/Regional-Initiatives/Africa-Clean-Energy-Corridor
IRENA. (2023). Word energy transitions outlook 2023. https://www.irena.org/-/media/Files/IRENA/Agency/Publication/2023/Jun/IRENA_World_energy_transitions_outlook_2023.pdf?rev=db3ca01ecb4a4ef8accb31d017934e97
Jenkins, K., McCauley, D., Heffron, R., Stephan, H., & Rehner, R. (2016). Energy justice: A conceptual review. *Energy Research and Social Science, 11*, 174–182. https://doi.org/10.1016/j.erss.2015.10.004

Jenkins, K. E. H., Sovacool, B. K., Mouter, N., Hacking, N., Burns, M. K., & McCauley, D. (2021). The methodologies, geographies, and technologies of energy justice: A systematic and comprehensive review. *Environmental Research Letters, 16*, 043009.

Jerez, B., Garcés, I., & Torres, R. (2021). Lithium extractivism and water injustices in the Salar de Atacama, Chile: The colonial shadow of green electromobility. *Political Geography, 87*.https://doi.org/10.1016/j.polgeo.2021.102382

Kanschik, P. (2016). Eco-sufficiency and distributive sufficientarianism—friends or foes? *Environmental Values, 25*(5), 553–571.

Katz-Lavigne, S. (2019). Artisanal copper mining and conflict at the intersection of property rights and corporate strategies in the Democratic Republic of Congo. *The Extractive Industries and Society, 6*(2). https://doi.org/10.1016/j.exis.2018.12.001

Kirshner, J., & Power, M. (2015). Mining and extractive urbanism: Postdevelopment in a Mozambican boomtown. *Geoforum, 61*, 67–78. https://doi.org/10.1016/j.geoforum.2015.02.019

Knierzinger, J. (2018). *Bauxite mining in Africa—Transnational corporate governance and development*. Palgrave Macmillan.

Lacey-Barnacle, M. (2020). Proximities of energy justice: contesting community energy and austerity in England. *Energy Resourcing Social Science, 69*.https://doi.org/10.1016/j.erss.2020.101713

Lehtonen, T., Heikkurinen, P. (2022). Sufficiency and sustainability: conceptual analysis and ethical considerations for sustainable organisation. *Environmental Values, 31*(1), 599–618.

Lennon, B., Dunphy, N. P., & Sanvicente, E. (2019). Community acceptability and the energy transition: a citizens' perspective. *Energy Sustainable Society, 9*(35). https://doi.org/10.1186/s13705-019-0218-z

Liu, W., & Agusdinata, D. B. (2020). Interdependencies of lithium mining and communities sustainability in Salar de Atacama, Chile. *Journal of Cleaner Production, 260*.https://doi.org/10.1016/j.jclepro.2020.120838

Malerba, D. (2022). Just transitions: A review of how to decarbonise energy systems while addressing poverty and inequality reduction.

Manno, J. (2017). Environmental sustainability/sufficiency. In R. A. Denemark & R. Marlin-Bennett (Eds.), *The International Studies Encyclopedia*. Wiley-Blackwell.

McCauley, D., & Heffron, R. (2018). Just transition: Integrating climate, energy and environmental justice. *Energy Policy, 119*, 1–7. https://doi.org/10.1016/j.enpol.2018.04.014

McCauley, D., Heffron, R., Stephan, H., & Jenkins, K. (2013). Advancing energy justice: The triumvirate of tenets and systems thinking. *International Energy Law Review, 32*(3), 1–5.

McCauley, D., Ramasar, V., Heffron, R. J., Sovacool, B. K., Mebratu, D., & Mundaca, L. (2019). Energy justice in the transition to low carbon energy systems: Exploring key themes in interdisciplinary research. *Applied Energy, 233–234*, 916–921. https://doi.org/10.1016/J.APENERGY.2018.10.005

Milchram, C., Hillerbrand, R., van de Kaa, G., Doorn, N., & Künneke, R. (2018). Energy justice and smart grid systems: Evidence from the Netherlands and the United Kingdom. *Applied Energy, 229*, 1244–1259. https://doi.org/10.1016/j.apenergy.2018.08.053

Milchram, C., Künneke, R., Doorn, N., van de Kaa, G., & Hillerbrand, R. (2020). Designing for justice in electricity systems: A comparison of smart grid experiments in the Netherlands. *Energy Policy, 147*, 111720. https://doi.org/10.1016/j.enpol.2020.111720

Mitchell, S. (2023). *Energy poverty: A hidden crisis*. Brighton, England

Mongabay. (2024). Indigenous communities make clean energy drive work for, not against, them. https://news.mongabay.com/2024/06/indigenous-communities-make-clean-energy-drive-work-for-not-against-them/

Morello-Frosch, R., & Obasogie, O. K. (2023). The climate gap and the color line—Racial health inequities and climate change. *New England Journal of Medicine, 388*(10), 943–949. https://doi.org/10.1056/nejmsb2213250

National Conference of State Legislatures (NCSL). (2022). Energy Justice and the Energy Transition. https://www.ncsl.org/energy/energy-justice-and-the-energy-transition

Nations, U. (2014). World survey on the role of women in development 2014: Gender equality and sustainable development.
Nem Singh, J. T., & Camba, A. A. (2016). Neoliberalism, resource governance and the everyday politics of protests in the Philippines. In *The everyday political economy of Southeast Asia* (pp. 49–71). Cambridge University Press.
Nickel, N. (2023). Supporting the UN sustainable development goals-air. https://nornickel.ru/sustainability/environment/air/
Office of Community Services. (2024). Low income home energy assistance program (LIHEAP). https://www.acf.hhs.gov/ocs/programs/liheap
Ould Amrouche, S., Rekioua, D., Rekioua, T., & Bacha, S. (2016). Overview of energy storage in renewable energy systems. *International Journal of Hydrogen Energy, 41*(45), 20914–20927. https://doi.org/10.1016/j.ijhydene.2016.06.243
Oyu Tolgoi Watch. (2012). A useless Sham—A Review of the Oyu Tolgoi Copper/Gold Mine Environmental and Social Impact Assessment.
Pager, D., & Shepherd, H. (2008). The sociology of discrimination: Racial discrimination in employment, housing, credit, and consumer markets. *Annual Review of Sociology, 34*, 181–209. https://doi.org/10.1146/annurev.soc.33.040406.131740
Papaefthymiou, G., & Dragoon, K. (2016). Towards 100% renewable energy systems: Uncapping power system flexibility. *Energy Policy, 92*, 69–82. https://doi.org/10.1016/j.enpol.2016.01.025
Princen, T. (2003). Principles for sustainability: From cooperation and efficiency to sufficiency. *Glob Environ Polit, 3*(1), 33–50. https://doi.org/10.1162/152638003763336374
Radwin, M. (2024). Rapid growth of Bolivia's lithium industry creating new problems for local communities.
Raworth, K. (2017). *Doughnut economics : Seven ways to think like a 21st-century economist*. Cornerstone.
Reglan Mine. (2021). Reglan Mine-Our engagement.
Reuters. (2021). Greenland minerals won't attend public meeting over rare earths project.
Rosyida, I., Khan, W., & Sasaoka, M. (2018). Marginalization of a coastal resource-dependent community: A study on tin mining in Indonesia. *The Extractive Industries and Society, 5*(1). https://doi.org/10.1016/j.exis.2017.11.002
Saes, B. M., Del, B. D., Neyra, R., Wagner, L., & Martínez-Alier, J. (2021). Environmental justice and corporate social irresponsibility: The case of the mining company Vale S.A. Ambiente e Sociedade, 24. https://doi.org/10.1590/1809-4422asoc20210014vu2021L4ID
Sagárnaga, R. (2015). Bolivia's lithium boom: dream or nightmare?.
Sairinen, R., Tiainen, H., & Mononen, T. (2017). Talvivaara mine and water pollution: An analysis of mining conflict in Finland. *The Extractive Industries and Society, 4*(3). https://doi.org/10.1016/j.exis.2017.05.001.
Salleh, A. (2009). From eco-sufficiency to global justice. In A. Salleh (Ed.), *Eco-sufficiency and global justice: women write political ecology* (pp. 407–421). Pluto Press.
Sanchez-Lopez, M. D. (2021). Territory and lithium extraction: The Great Land of Lipez and the Uyuni Salt Flat in Bolivia. *Political Geography, 90*.https://doi.org/10.1016/j.polgeo.2021.102456
Schlosberg, D. (2004). Reconceiving environmental justice: Global movements and political theories. *Environmental Politics, 13*(3), 517–540. https://doi.org/10.1080/0964401042000229025
Shakeel, S. R. (2019). Commercialization of renewable energy technologies : A study of socio-economic, technical and regulatory factors in Finland and Pakistan. Acta Wasaensia (Vol. 430, University of Vaasa).
Shakeel, S. R., Takala, J., & Zhu, L.-D. (2017). Commercialization of renewable energy technologies: A ladder building approach. *Renewable and Sustainable Energy Reviews, 78*, 855–867. https://doi.org/10.1016/j.rser.2017.05.005

Shang B. (2022). Tackling gender inequality: Definitions, trends, and policy designs. In: IMF Working Paper. https://www.elibrary.imf.org/view/journals/001/2022/232/001.2022.issue-232-en.xml

Siciliano, G., Urban, F., Tan-Mullins, M., & Mohan, G. (2018). Large dams, energy justice and the divergence between international, national and local developmental needs and priorities in the global South. *Energy Research and Social Science, 41*, 199–209. https://doi.org/10.1016/j.erss.2018.03.029

SIFEE. (2023). Research on the consideration of social impacts in impact assessment processes applicable to Nunavik.

Simon, F. (2020). Biggest challenge in energy transition is social justice. In: EURACTIV. https://www.euractiv.com/section/energy-environment/news/biggest-challenge-in-energy-transition-is-social-justice-eu-official-says/

Sovacool, B. K. (2019). The precarious political economy of cobalt: Balancing prosperity, poverty, and brutality in artisanal and industrial mining in the Democratic Republic of the Congo. *The Extractive Industries and Society, 6*(3). https://doi.org/10.1016/j.exis.2019.05.018.

Sovacool, B. K., Ali, S. H., Bazilian, M., Radley, B., Nemery, B., Okatz, J. & Mulvaney, D. (2020). Sustainable minerals and metals for a low-carbon future. *Science (80), 367*(6473), 30–33. https://doi.org/10.1126/science.aaz6003

Sovacool, B. K., Burke, M., Baker, L., Kotikalapudi, C. K., & Wlokas, H. (2017). New frontiers and conceptual frameworks for energy justice. *Energy Policy, 105*, 677–691. https://doi.org/10.1016/j.enpol.2017.03.005

Sovacool, B. K., & Dworkin, M. H. (2015). Energy justice: Conceptual insights and practical applications. *Applied Energy, 142*.https://doi.org/10.1016/j.apenergy.2015.01.002

Sun, Z., Zhang, F., Wang, Y., Shao, Z. (2023). Literature review and analysis of the social impact of a just energy transition. *Frontiers in Sustainable Food Systems, 7*, 1119877.

STUK. (2024). Talvivaaran kaivos.

TEM (2017) KORKEIMMAN HALLINTO-OIKEUDEN PÄÄTÖS.

Tiainen, H., Sairinen, R., & Mononen, T. (2014). Talvivaaran kaivoshankkeen konfliktoituminen [Conflict process of Talvivaara mine in Finland]. Ympäristöpolitiikan ja –oikeuden Vuosik VII:7–76.

Tinto R (2024) Rio Tinto- Communities. https://www.riotinto.com/en/sustainability/communities

Tolonen, E., Shakeel, S. R., & Juntunen, J. K. (2023). Promoting just transition or enhancing inequalities? Reflection on different energy community business models in terms of energy justice. In *Lecture Notes in Energy*.

U-tantada, S., Mujtaba, B. G., Yolles, M., & Shoosanuk, A. (2016). Sufficiency economy and sustainability. In *2nd Multidisciplinary Research and Innovation for Globally Sustainable Development* (pp. 1–8).

United Nations. (2021). Theme report on energy transition: Towards the achievement of SDG 7 and net-zero emissions. https://www.un.org/sites/un2.un.org/files/2021/11/2021-twg_2.pdf

United Nations Climate Change. (2024). Green climate fund. https://unfccc.int/process/bodies/funds-and-financial-entities/green-climate-fund?gad_source=1&gclid=Cj0KCQjwsJO4BhDoARIsADDv4vA_3LBw9O_vSSukwHkgCQphGp2ArA_bliEPxyURJUCasmmwdxwmP3MaAkMEEALw_wcB

Vale. (2021). Integrated report 2020.

Vale. (2022). Integrated Report 2021.

Valkhof, B. (2020). Energy transition 101: Getting back to basics for transitioning to a low-carbon economy. In World Economics. Forum. https://www3.weforum.org/docs/WEF_Energy_Transition_101_2020.pdf

Whaples, R. M., Michael, C. M., & Christopher J. C. (2023). Is Social Justice Just. Independent Institute, Oakland.

Wiegink, N. (2020). Learning lessons and curbing criticism: Legitimizing involuntary resettlement and extractive projects in Mozambique. *Political Geography, 81*.https://doi.org/10.1016/j.polgeo.2020.102192

Woodworth, M. D. (2017). Disposable Ordos: The making of an energy resource frontier in western China. *Geoforum, 78*.https://doi.org/10.1016/j.geoforum.2016.04.007

World Resources Institute. (2021). Spain's national strategy to transition coal-dependent communities. https://www.wri.org/update/spains-national-strategy-transition-coal-dependent-communities

Yakovlev, A. S., Plekhanova, I. O., Kudryashov, S. V., Aimaletdinov, R. A. (2008). Assessment and regulation of the ecological state of soils in the impact zone of mining and metallurgical enterprises of Norilsk Nickel company. *Eurasian Soil Science, 41*(6). https://doi.org/10.1134/S1064229308060100

Chapter 8
A Survey on Optimization Techniques and Modeling Tools in Sustainable Energy Planning

Theyab R. Alsenani and Mohammad Mehdi Amiri

Abstract The search for optimal methods and strategies to achieve efficient and environmentally friendly energy systems is at the core of optimization techniques in sustainable energy planning. This book chapter delves into the interdisciplinary field of sustainable energy planning, emphasizing the central role of optimization techniques in addressing the complexities of energy resource allocation, demand management and environmental impact reduction. The chapter explores mathematics and advanced algorithms in the context of energy infrastructure optimization tools, demonstrating the importance of their use in achieving sustainability goals. Real-world case studies and examples illustrate the practical application of these techniques, providing readers with valuable insights into the intricacies of designing and managing sustainable energy systems. As the global pursuit of clean energy intensifies, this chapter serves as a valuable resource for researchers, practitioners and policy makers seeking to optimize energy planning processes for a more sustainable and resilient future. In this chapter, the authors categorize and re-categorize optimization models in sustainable energy planning and review their optimization techniques. Also, to fully familiarize with the types of models presented, the latest and most important models have been reviewed and analyzed in line with the development of sustainable energy.

Keywords Sustainable energy · Renewable energy · Optimization models · Programming models · Game models · Multi-attribute decision models · Goal programming · Sustainable development goals (SDGs)

T. R. Alsenani
Department of Electrical Engineering, College of Engineering in Al-Kharj, Prince Sattam Bin Abdulaziz University, Al-Kharj, Saudi Arabia
e-mail: t.alsenani@psau.edu.sa

M. M. Amiri (✉)
Faculty of Electrical Engineering, Shahid Beheshti University, Tehran, Iran
e-mail: moh_amiri@sbu.ac.ir

I. Rahimi et al. (eds.), *Sustainable Supply Chain of Renewable Energy Networks*, Energy Systems in Electrical Engineering, https://doi.org/10.1007/978-981-96-4940-2_8

8.1 Sustainable Energy Planning Concept

The concept of sustainable development implies a higher level of social and environmental responsibility toward the planet and its resources. Increasing energy efficiency and switching to renewable energy sources (RESs) are essential to a better and more sustainable future. During this transition, decentralization increases and complex requirements for energy system planning arise. The focus will be on regenerative and partly decentralized technologies, since large-scale power plants are shifting toward innovative combinations of regenerative and partly decentralized technologies. The mathematical models produced by this process become considerably larger and more complex as the decision options become more closely intertwined, and, probably, they become impractical to solve using generic methods and standard solvers as the decision options become more complicated. A priority of sustainable development is the transition of the energy sector, which is concerned with managing financial, technological, institutional, natural, and social resources for current and future generations, as well as mitigating global warming, which constitute key areas of importance for energy planning optimization (Radovanović et al., 2021).

Sustainable Development Goals (SDGs) are a set of goals introduced by the United Nations in 2015 (Tsangas et al., 2022). Three key targets are included in SDG 7, namely, ensuring affordable and reliable access to modern energy for all, increasing the share of renewable energy in the energy mix worldwide, and improving the rate of global energy efficiency improvement (Gielen et al., 2019). In addition, SDG 11 aims to create sustainable cities and communities, which correspond to responsible consumption and production (SDG 12) (Doni et al., 2020; Swain & Karimu, 2019; Tsangas et al., 2022). The SDGs have been used to create the European Green Deal (EGD), which aims to make Europe carbon neutral by 2050. More than 75% of the European Union's (EU) greenhouse gas emissions come from the production and consumption of energy (European Commission, 2019). To meet the climate goals of the EGD, decarbonizing the EU's energy system is therefore essential. For the energy sector, the EGD emphasizes the following three key principles in order to achieve a transition to clean energy, all of which will contribute to reducing greenhouse gas emissions and improving the quality of life for our citizens:

- Maintaining an affordable and stable energy supply for the EU,
- The development of a fully integrated and interconnected EU energy market, focusing on energy efficiency and building efficiency improvement,
- Establishing a power sector-based primarily on renewable resources.

With major lifestyle changes in developed countries, close cooperation among countries, and several measures, it is possible to replace fossil fuels with renewable energy by 2050. Renewable energy is also essential, along with improving energy efficiency, conserving energy, and taxing carbon, balancing human well-being with energy consumption per capita, implementing cap-and-trade systems, capturing, utilizing, and storing carbon, and developing nuclear power. A number of these actions affect society, the economy, and the environment, and are also in line with SGD 7 and SDG 13 (climate action).

8.2 Energy Planning Challenges and the Role of Optimization Models

As part of the development of energy systems, optimization models play an important role in addressing several challenges. Energy systems can be developed more sustainably, reliably, and efficiently with these models, which address both technical and economic challenges. Here are some key challenges and how optimization models can help tackle them (Amiri et al., 2024; Kotzur et al., 2021; Scherer et al., 2018):

- Renewable energy integration

RESs are inherently intermittent and weather-dependent which makes it difficult for power grids to stabilize supply and demand. To integrate these resources efficiently, optimization models play an important role in the design and management of energy systems. In this context, advanced optimization techniques incorporate uncertainty arising from the variable nature of RESs. Stochastic programming, robust optimization, and chance-constrained models are commonly employed to handle this uncertainty. These approaches account for variations in renewable energy output by modeling uncertain parameters (such as solar irradiance or wind speed) as probabilistic distributions or uncertain sets, allowing for the design of energy systems that are not only optimized for average conditions but also resilient to fluctuations. By doing so, they ensure a stable and reliable energy supply, even under extreme or unforeseen conditions, enhancing the overall efficiency and security of renewable energy integration (Kotzur et al., 2021; Scherer et al., 2018).

- Energy storage management

Managing energy storage and demand based on supply and demand and efficiently taking advantage of RESs is a challenging problem. Energy storage systems (ESS) are therefore used to solve the variability by enabling any excess energy produced at one time to be stored for use at various low energy production times. Indeed, optimization models can be considered as decision-making instruments for defining the proper size, location, and operational strategies for ESS to make maximum profit. Of all the storage technologies available, hydrogen storage has very high energy density, can store several forms of energies for long durations of time. Hydrogen by water electrolysis is possible when renewable energy is excess, can be stored for later use or converted into other applications power including transport and manufacturing. It lies within the framework of hydrogen storage to consider the impacts of hydrogen infrastructure on such planning decisions and the virtually unlimited large-scale storage option inherent to hydrogen systems (Amiri et al., 2024). These models help ensure that hydrogen storage complements other energy storage technologies, contributing to a more efficient energy system that minimizes both costs and carbon emissions.

- Economic and environmental trade-offs

It is quite difficult to optimize the economic cost of a system, while minimizing the environmental effects which is quite crucial when transitioning to clean systems. In this context, multi-objective optimization models play a significant role in comparing the various options and the consequence of their implementation by advancing the policy maker and the stakeholders to take informed decisions that have regard to the economic and the environmental objectives. For instance, although hydrogen has been widely advocated as a zero-emission carbon-free fuel for heavy industry and transport that cannot easily be electrified, this is disingenuous in a context of resource consumption. But, the main challenge hindering the use of hydrogen is the high costs of generation, storage, and transportation. Economic evaluation of integration of hydrogen technologies shall use optimization models to evaluate costs associated with integration of hydrogen technologies and the benefits that may be accrued from reduced emission of greenhouse gases (Amiri et al., 2024). These models, by including hydrogen as one of the variables of analysis, measure the financial commitments needed for infrastructure construction against the long-term environmental benefits, providing policy makers with a realistic view of energy system planning.

- Capacity expansion planning

Planning for future energy infrastructure investments is a complex and multifaceted task, particularly due to uncertainties in demand growth, technological advancements, fuel price volatility, and policy changes aimed at decarbonization. Capacity expansion planning entails the identification of the suitable investments necessary to enhance capacity for generation, transmission, and storage in the future to reassume the electricity needs. These forms of future investment planning rely on enhanced optimization models as these techniques help to model the different possible working scenarios based on demand uncertainty, renewable energy connection, fuel prices, and regulations. These models may use mixed integer linear programming (MILP) or stochastic programming since capacity additions are discrete and the evolution of some parameters is stochastic (Scherer et al., 2018).

Introduction of the RESs and new technologies such as hydrogen production and storage makes planning challenging again due to the fact that these technologies must take into account intermittency, storage needs, and cost reductions due to learning effect. Optimization models provide insights into the optimal timing, scale, and technology mix for capacity expansions, considering not only capital expenditures but also operating costs, carbon pricing, and system reliability. For example, they help determine when to invest in grid-scale battery storage, hydrogen storage systems, or new renewable capacity to minimize costs while meeting emissions targets. These models also enable planners to explore scenarios under different policy conditions, such as carbon taxes or renewable portfolio standards, helping stakeholders make informed decisions on the most sustainable pathways for future energy infrastructure development.

- Scalability of optimization models

A critical challenge in sustainable energy planning is the scalability of optimization models, particularly when applied to large, complex systems such as regional or national energy networks. With the increase of energy systems' scale involving generation units, transmission lines, storage stations, and RESs, the number of decision variables and constraints in the models escalates. This brings into computation difficulty and it becomes challenging to solve problems of large samples within reasonable time and space using standard methods.

Computational complexity is the measure of what amount of computations in terms of time and space needed in order to solve a certain problem with growing size of the input data. The optimization models may include mixed integer decision variables, non-linear equations, a stochastic approach to model the variability of renewable energy generation, demand and fuel prices in energy systems. These factors make the underlying models computer intensive and practically insolvable when planning for multi-year capacity addition or real-time operations in large geographical areas. LP and MIP, decomposition, metaheuristic, stochastic, and robust optimization have different scalability and while they may be efficient on large systems, their efficiency differs depending on the context of the current situation.

8.3 Optimization Models in Sustainable Energy Planning

Optimization models are mathematical tools that help find the best solutions for a given problem or objective, such as minimizing costs, maximizing efficiency, or reducing emissions. Optimization models can be useful for planning and designing sustainable energy systems, which rely on renewable and low-carbon sources of energy (Klemm & Wiese, 2022). There are many types of optimization models, each with different characteristics, advantages, and disadvantages (Yu et al., 2023). In this book chapter, we consider optimization models including the following four categories, and for each of them, the most important types will be introduced.

- Programming models

These models use mathematical expressions to represent the objective function and the constraints of the problem. Programming models can be classified into different categories, such as linear, non-linear, integer or mixed-integer programming, depending on the nature of the variables and functions involved.

- Multi-attribute decision models

These models deal with problems involving multiple and conflicting criteria or objectives, such as minimizing cost and emissions, or maximizing efficiency and reliability. Multi-objective decision models can be classified into different categories, such as multi-objective optimization, objective programming, or analytic hierarchy process, depending on the method of aggregating or ranking the objectives.

- Game models

These models use game theory to analyze the strategic interactions and decisions of multiple agents or players, such as energy producers, consumers, regulators or competitors. Game models can be classified into different categories, such as cooperative, non-cooperative or evolutionary games, depending on the assumptions and rules of the game.

- Hybrid models

These models combine two or more of the above models or other methods, such as simulation, forecasting or evaluation, to produce a more comprehensive and realistic model of the problem. Hybrid models can be classified into different categories, such as stochastic, robust, or fuzzy optimization, depending on how they deal with uncertainty, variability or imprecision in the input data or model parameters.

8.3.1 Programming Models

The most important categories of programming models are as follows:

- Linear programming (LP): Optimization where both the objective function and constraints are linear (Rizzo & Savino, 2012).
- Nonlinear programming (NLP): Involves non-linear objective functions (OF) or constraints (Elkasrawy et al., 2024; Fan et al., 2022). For example, this problem due to its OF is known as a NLP.

$$\begin{array}{ll} \text{OF} & z = 20x1 + 30x_2 - 0.5x_1^2 \\ & 2x_1 + 3x_2 \leq 100 \\ \text{Constraints} & x_1 + 2x_2 \leq 80 \\ & x_1, x_2 \geq 0 \end{array} \tag{8.1}$$

- Mixed-integer programming (MIP): Combines integer and continuous variables (Cosic et al., 2021; Jin et al., 2016).
- Dynamic programming (DP): DP breaks a problem down into simpler subproblems, solving them recursively. A classic example is the Knapsack problem.

Example Given a knapsack with a weight capacity of 5 units, maximize the value of the items you can carry.

- Item 1: Value = 10, Weight = 3
- Item 2: Value = 6, Weight = 2
- Item 3: Value = 8, Weight = 4.

Formulation: Let $V(i,w)$ be the maximum value attainable with the first i items and a knapsack of weight w. Recurrence relation is as follows:

$$v(i, w) = \begin{cases} V(i-1, w) & \text{if } \mathrm{w}_i > w \\ \max(V(i-1, w), V(i-1, w - w_i + v_i) & \text{if } \mathrm{w}_i \le w \end{cases} \tag{8.2}$$

where w_i is the weight of item i and v_i _ivi is the value of item i.

8.3.2 *Multi-attribute Decision Models*

I. *Multi-objective optimization*

Multi-objective optimization is a process in optimization problems where multiple, often conflicting objectives are simultaneously optimized. Instead of finding a single optimal solution, this approach seeks to identify a set of Pareto-optimal solutions, where no objective can be improved without degrading another (Alshammari et al., 2023; Ervural et al., 2018). Two important examples of these models are as follows:

- Weighted sum method
- ε-Constraint method (Balaman, 2016).

General problem formulation for both method is as (8.3):

$$\begin{array}{ll} \text{OF} & \min_x\{f_1(x), f_2(x)\} \\ \text{Subject to constraints:} & g_i(x) \le 0, \quad \mathrm{i} = 1, 2, \ldots, \mathrm{m} \end{array} \tag{8.3}$$

where $g_i(x)$ are inequality constraints.

In the weighted sum method, multiple objectives are combined into a single objective function by assigning weights to each objective. The problem then becomes a single objective optimization problem. Weighted sum formulation is as follows:

$$\begin{array}{ll} \text{OF} & \min_x [\lambda_1 f_1(x) + \lambda_2 f_2(x)] \\ \text{Subject to constraints:} & g_i(x) \le 0, \quad \mathrm{i} = 1, 2, \ldots, \mathrm{m} \end{array} \tag{8.4}$$

where λ_1 and λ_2 are the weights for each objective function, such that $\lambda_1 + \lambda_2 = 1$ and $\lambda_1, \lambda_2 \ge 0$. By changing the values of λ_1 and λ_2, we can explore different trade-offs between the objectives. This method is simple but may not always find solutions in non-convex regions of the Pareto front.

In the ε-constraint method, we optimize one objective, say $f_1(x)$ and transform the other objective $f_2(x)$ into a constraint (8.5):

$$\begin{array}{ll} \text{OF} & \min_{x} f_1(x) \\ \text{Subject to} & f_2(x) \leq \varepsilon \\ \text{constraints:} & g_i(x) \leq 0, \quad i = 1, 2, \ldots, m \end{array} \tag{8.5}$$

where ε is a specified upper bound for the second objective $f_2(x)$. By varying ε, we can explore different trade-offs between the objectives.

II. *Goal programming*

Goal programming is a branch of multi-objective optimization where the aim is to achieve multiple goals or objectives simultaneously. Unlike traditional optimization, which seeks to optimize a single objective, GP allows for the setting of priority levels or weights for different goals. This is particularly useful when dealing with conflicting objectives, as it enables the decision-maker to find a solution that best satisfies the prioritized goals (Bakhtavar et al., 2020). In the context of power systems and energy planning, GP is often used to balance trade-offs between economic, environmental, and technical objectives, facilitating more comprehensive and practical decision-making (Jayaraman et al., 2017a, 2017b). The following are the most important types of this programming.

- Weighted goal programming
- Lexicographic goal programming
- Chebyshev goal programming.

III. *Analytic hierarchy process (AHP)*

AHP involves decomposing a complex decision problem into a hierarchy of more easily comprehended sub-problems, each of which can be analyzed independently. By assigning weights to each criterion based on pairwise comparisons, AHP facilitates the aggregation of these weights to derive a ranked list of alternatives or solutions (Chiu et al., 2021; Zaheb et al., 2024).

IV. *Analytic network process (ANP)*

Unlike AHP, which assumes a hierarchical structure with independent criteria, ANP allows for more complex, network-like relationships where criteria can influence each other in a feedback loop. This method is particularly useful in power systems and energy planning, where various factors—such as economic, technical, environmental, and social considerations—are often interrelated (Kheybari et al., 2020).

8.3.3 Game Models

I. *Cooperative game theory*

Cooperative game theory in optimization problems focuses on how groups of agents or decision-makers can collaborate to achieve a collective goal, sharing the benefits of their cooperation. In this framework, the participants form coalitions and work together to optimize a common objective, such as minimizing costs or maximizing efficiency. The theory provides mechanisms for fairly distributing the collective gains among the participants, ensuring that each party benefits from the collaboration (Churkin et al., 2021; Kong et al., 2022).

II. *Non-cooperative game theory*

Non-cooperative game theory in optimization problems examines situations where multiple independent agents or decision-makers pursue their own interests, often in competition with each other. Unlike cooperative game theory, there is no collaboration between the players; instead, each one seeks to optimize their own objective, considering the strategies and potential actions of others. Two important types of these models are as follows:

- Nash equilibrium (Pfeifer et al., 2023)
- Stackelberg equilibrium (Ma et al., 2024; Shi et al., 2020).

III. *Evolutionary game theory*

Evolutionary game theory in optimization problems explores how strategies evolve over time among competing agents in a dynamic environment. Unlike traditional game theory, which assumes rational players with fixed strategies, evolutionary game theory models how strategies change and adapt based on their success or failure within a population. This approach is particularly relevant in complex systems like power grids and energy markets, where multiple agents interact repeatedly, and strategies may evolve in response to environmental changes, technological advancements, or regulatory shifts. Evolutionary game theory provides insights into the long-term stability and adaptability of strategies, helping to optimize decision-making processes in scenarios where conditions and behaviors are continually evolving (Wang et al., 2022).

8.3.4 Hybrid Models

Main classifications of hybrid models are as follows:

- Hybrid metaheuristics: Combines metaheuristic algorithms like genetic algorithms with other optimization techniques (Güven & Samy, 2022).

- Hybrid simulation–optimization: Integrates simulation models with optimization methods (Figueira & Almada-Lobo, 2014).
- Hybrid machine learning (ML) models: Uses ML algorithms in conjunction with optimization techniques (Sharmila et al., 2019).

8.4 Trends in Energy Sustainability Planning Assessment

Over the past few years, sustainable energy planning and relevant research have increasingly focused on achieving SGD 7 and SDG 13. In order to develop energy strategies and sustainable energy criteria and indicators, society, the economy, and, especially, the environment are all taken into account, with global warming being the main concern. Here are some trends in energy sustainability planning assessment. These trends highlight the complex and multifaceted nature of modern energy sustainability planning, with an emphasis on integrating diverse criteria and methodologies to meet global sustainability goals.

- Application of sustainability indicators: Performance indicators quantifying sustainability concern in energy planning are increasingly being applied, generally around environmental, social, and economic factors (Adeyemi-Kayode et al., 2022; Mehmood et al., 2022).
- Global warming potential priority: Particularly at this modern age, the environmental concern about global warming potential has become a key consideration in energy sustainability planning.
- Multi-criteria decision analysis (MCDA) adoption: It includes MCDA methods, such as AHP, TOPSIS, VIKOR, and PROMETHEE that have been widely used to evaluate and rank energy strategies with regards to multiple criteria, including technical, economic, environmental, and social dimensions (Adeyemi-Kayode et al., 2022; Genco & Genco, 2021; Papapostolou et al., 2019; Tutak et al., 2020).
- Diversification of criteria: In recent times, the assessment criteria of sustainability have increased to include other wide-ranging impacts—greenhouse gas emissions, energy use, land use, social acceptance, safety risks, among others (Klemm & Wiese, 2022).
- Regional/sectoral studies: Many of the studies focus on specific regions/sectors, such as the energy mix in Pakistan, power technology in Kenya, and energy security in the Baltic states (Bilal et al., 2022; Manirambona et al., 2022; Streimikiene, 2020).
- Composite indicators development: new indices are being developed, like the Sustainable Energy Development Aggregated Index, and the Energy Transition and Sustainable Development Index, supporting the assessment and ranking of countries or regions with respect to their progress toward sustainable energy.
- Attention to regional specificities: Sustainability assessments are adapted to regional contexts-for example, the energy strategy for Iran, New England Normandy, and Baltic states-reflecting environmental, social, and economic conditions of particular regions (Streimikiene, 2020).

- Hybrid methods: Combined methods such as AHP–TOPSIS and hybrid AHP–TOPSIS have been used to address complicated decision-making procedures in energy planning (Genco & Genco, 2021; Tutak et al., 2020).
- Biodiversity focus: There is also growing attention to address biodiversity loss fairly and equitably, alongside climate change (Key 2024 Sustainability Trends Driving the Year Ahead, 2024).
- Decarbonization Efforts: There is currently a strong drive for the decarbonization of the economy, and much investment in RESs such as wind and solar (Penna et al., 2023).

8.5 Modeling Tools for Sustainable Development

As mentioned earlier, the researchers have presented many models in order to evaluate and achieve sustainable energy development in the field of optimization. Figure 8.1 provides an overview of these models. In this section, we have reviewed 20 of the most important of these models and they have been briefly described. The basis for choosing and explaining these 20 models is their novelty in the recent studies of researchers.

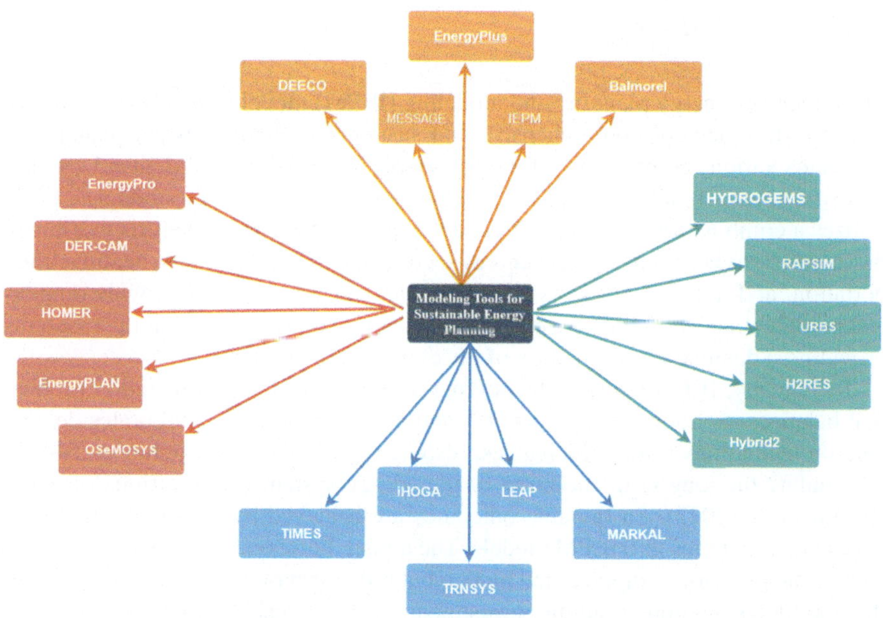

Fig. 8.1 20 important modeling tools for sustainable development

8.5.1 MESSAGE Model

Model for Energy Supply Strategy Alternatives and their General Environmental Impact (MESSAGE) is used for planning medium or long-term energy systems, analyzing climate change policies, and developing scenarios. It is designed to evaluate and formulate long-term energy strategies by simulating the development of energy supply, conversion, and utilization technologies under various economic, technical, and environmental constraints. MESSAGE employs a linear programming framework to minimize the total system cost while meeting specified energy demand and environmental targets, making it particularly effective for sustainable energy planning. A time-step of 5 or 10 years is used to simulate a maximum of 120 years in this model. MESSAGE can simulate all thermal generation, renewable energy, storage, and conversion technologies as well as carbon sequestration. Model inputs are very detailed on the supply side, but more aggregated on the demand side. Climate targets are not estimated by the model, but rather global, regional and multisector mitigation strategies. With MESSAGE, it is possible to determine cost-effective portfolios of greenhouse gases emission limitation and reduction measures (Huppmann et al., 2019).

8.5.2 IEPM Model

The integrated energy planning and modeling (IEPM) model is an advanced analytical tool designed for comprehensive and sustainable energy system planning. It integrates various components of energy supply, conversion, and demand within a unified framework, allowing for the holistic assessment of energy strategies. IEPM utilizes a combination of optimization techniques and scenario analysis to evaluate the interactions between different energy sectors, including electricity, heating, transportation, and industrial processes. This model can accommodate a wide range of technologies, such as RESs, fossil fuels, nuclear power, and energy storage systems, providing a detailed representation of the entire energy system.

Technically, IEPM operates by solving a multi-objective optimization problem that balances economic, environmental, and social criteria. It incorporates detailed data on technology costs, efficiencies, resource availability, and policy constraints to simulate the long-term evolution of the energy system under various scenarios (Bashir et al., 2022). Sustainable outcomes are achieved by incorporating energy system uncertainties into IEPM models. The model, however, does not rely solely on a one-time plan or a computational model, but rather is based on a dynamic structure that facilitates ongoing planning, investment, decision-making, and execution. In a model such as this, stakeholders are committed and communicated, future risks and uncertainties are evaluated, short-term actions are aligned with long-term plans, and alternative solutions are analyzed.

8.5.3 Balmorel Model

The Balmorel model is a widely used optimization tool in sustainable energy planning. This model is a sophisticated, open-source energy system model designed to support sustainable energy planning and policy analysis at both national and regional levels. It is specifically tailored for analyzing the integration of RESs into electricity and heat markets. Balmorel employs a linear programming framework to optimize the dispatch and investment in generation, transmission, and storage technologies under various economic, technical, and policy constraints. This flexibility allows it to simulate a wide range of scenarios, including different market designs, policy interventions, and technology developments, making it a valuable tool for evaluating the impacts of renewable energy integration on system reliability, costs, and emissions. Balmorel's model offers detailed spatial and temporal resolutions, enabling comprehensive analysis of both short-term operational decisions and long-term investment strategies. (Wiese et al., 2018; Yazdanie & Orehounig, 2023). Its applications in sustainable energy planning are as follows:

- Renewable energy integration: Balmorel is used to study the integration of RESs into the grid, assessing the impacts on system reliability, cost, and emissions.
- Capacity expansion: The model helps in planning future capacity expansions by determining the optimal mix of generation technologies to meet demand growth and policy goals.
- Transmission planning: Balmorel analyzes the need for new transmission infrastructure to facilitate the efficient flow of electricity, especially in regions with high renewable energy potential.
- Market analysis: It evaluates the functioning of electricity markets, including price formation, market power, and the impacts of market design changes.
- Scenario analysis: The model supports scenario analysis to explore different future energy pathways, considering various assumptions about technology costs, fuel prices, and policy frameworks.
- Policy impact assessment: Balmorel is used by governments and organizations to assess the impacts of energy policies on system costs, emissions, and energy security.

8.5.4 DEECO Model

The dynamic energy, emissions, and cost optimization (DEECO) model is a comprehensive framework designed to optimize the integration of decentralized energy resources within power systems, focusing on enhancing energy efficiency and sustainability. DEECO employs advanced optimization algorithms and communication protocols to manage the generation, storage, and consumption of energy at a local level, thereby reducing reliance on centralized power plants and enhancing the resilience of the energy system. The model facilitates the coordination of distributed energy resources (DERs) such as solar panels, wind turbines, batteries, and electric vehicles, ensuring that these resources operate synergistically to meet local energy demands while minimizing costs and environmental impacts.

In order to respond dynamically to changing market and grid conditions, the DEECO model implements real-time communication and control strategies. This capability enables the model to optimize the performance of micro-grids and smart grids, enhancing their ability to integrate RESs and improve energy efficiency. DEECO also supports the development of sustainable energy solutions that are both economically viable and environmentally friendly, contributing to the transition toward a more distributed and resilient energy infrastructure (Laveneziana et al., 2023).

8.5.5 EnergyPlus Model

EnergyPlus is a highly advanced building energy simulation software developed by the U.S. Department of Energy. It is designed to model energy consumption for heating, cooling, ventilation, lighting, and other energy flows in buildings. It is widely used in sustainable energy planning to evaluate and optimize the energy performance of buildings, enabling the design of energy-efficient buildings and retrofitting strategies that reduce energy use and environmental impact. EnergyPlus uses a detailed, physics-based approach to simulate the thermal and electrical dynamics of building systems, taking into account factors such as weather conditions, building materials, occupancy patterns, and HVAC systems.

EnergyPlus is one of the most flexible simulation tools due to its capability of describing complex interactions between the building components and their functional systems and, therefore, to show the impact of different options on the energy consumption. It is capable of accepting fine grained descriptions of building geometry and both construction and cladding materials, and can model the impact of different energy saving measures, and renewable electricity generation technologies such as solar photovoltaic and storage. EnergyPlus assists architects, engineers, and policymakers in reaching better decisions on making buildings more sustainable, efficient, and responsible for the generation of minimized greenhouse gases due to high energy uses, and support the general goals of sustainable development in buildings (Huang, 2024; Rezaei et al., 2021). Its most important applications in sustainable energy planning are as follows:

- Building design optimization
- Retrofitting existing buildings
- Net-zero energy buildings
- Renewable energy planning
- Policy and code compliance
- Demand response and load management
- Indoor environmental quality.

8.5.6 HYDROGEMS

The HYDROGEN Energy Model for Sustainability (HYDROGEMS) is designed to analyze and optimize the hydrogen energy supply chain, supporting the transition to a hydrogen-based energy system. It aims to facilitate sustainable energy planning by evaluating the technical, economic, and environmental aspects of hydrogen production, storage, and distribution. This model focuses on the integration of hydrogen technologies with RESs, such as solar and biomass, to facilitate the transition toward a sustainable and low-carbon energy future. By employing advanced optimization algorithms, HYDROGEMS can simulate various pathways for hydrogen production, including electrolysis, steam methane reforming, and biomass gasification, and assess their economic viability, environmental impact, and operational efficiency (Beuzekom et al., 2015).

This model can evaluate the interactions between hydrogen production and other energy sectors, such as transportation and industrial processes, providing a holistic view of the role of hydrogen in the energy system. By offering insights into optimal strategies for hydrogen integration, HYDROGEMS supports informed decision-making in sustainable energy planning, helping to identify cost-effective and environmentally friendly pathways for achieving long-term energy sustainability and resilience.

8.5.7 RAPSIM

Renewable Energy and Advanced Power System Integration Model (RAPSIM) is a comprehensive simulation tool designed for sustainable energy planning, specifically focused on the integration and optimization of RESs within power systems. It employs advanced modeling techniques to analyze the dynamic behavior of renewable energy technologies such as solar, wind, and hydroelectric power in conjunction with conventional power generation and storage systems (Jdeed et al., 2018). RAPSIM facilitates the assessment of system reliability, grid stability, and economic performance under various scenarios, including different levels of renewable penetration and grid configurations. The model supports decision-making processes by providing insights into optimal investment strategies, policy impacts, and operational efficiencies, ultimately aiming to enhance the sustainability and resilience of modern power systems (Khan & Minai, 2024).

8.5.8 URBS

The URBS model is an advanced linear optimization tool designed for sustainable energy planning, focusing on the optimal allocation and utilization of energy resources across multiple regions and time periods. It simulates the economic and environmental impacts of various energy supply and demand scenarios, incorporating RESs, storage systems, and transmission infrastructure. By considering constraints such as capacity limits, operational costs, and emissions, URBS provides a comprehensive framework for analyzing the trade-offs and synergies between different energy technologies and strategies. The model supports decision-making in energy policy and investment by offering detailed insights into cost-effective and sustainable pathways for energy system development, enhancing the integration of renewable energy and improving overall system efficiency and resilience (Höffner & Glombik, 2024).

8.5.9 H2RES

The Hydrogen Renewable Energy System (H2RES) model is a sophisticated simulation tool designed for the integrated analysis and optimization of hydrogen production and utilization within renewable energy systems. It focuses on the coupling of hydrogen technologies with various RESs to evaluate the potential for hydrogen to serve as a flexible energy carrier and storage solution. The model employs a mixed integer linear programming (MILP) framework to simulate the entire hydrogen supply chain, including production methods (e.g., electrolysis, steam methane reforming), storage options (e.g., compressed hydrogen, liquefied hydrogen), transportation logistics, and end-use applications across multiple sectors, such as transportation and industrial processes (Herc et al., 2021).

H2RES enables users to assess the economic viability and environmental impacts of different hydrogen strategies by providing insights into cost optimization, emissions reduction, and system reliability under various scenarios and operational conditions. By facilitating a deeper understanding of the role of hydrogen in achieving a sustainable energy future, H2RES plays a crucial role in guiding the transition toward low-carbon energy systems that enhance energy security and promote the integration of renewable resources (Herc et al., 2024).

8.5.10 Hybrid2

The Hybrid2 model is a comprehensive simulation tool designed for the optimal planning and integration of hybrid energy systems that combine RESs with conventional generation technologies and energy storage solutions (Rosso-Cerón et al., 2021). This model specifically focuses on evaluating the interactions and synergies between diverse energy generation technologies, such as solar photovoltaics (PV), wind, biomass, and fossil fuels, along with energy storage systems like batteries and pumped hydro storage. Utilizing advanced optimization techniques, Hybrid2 assesses various configurations and operational strategies to minimize costs, enhance reliability, and reduce greenhouse gas emissions, while also considering the technical and economic constraints of the energy system (Babatunde et al., 2020).

8.5.11 iHOGA

The Improved Hybrid Optimization by Genetic Algorithms (iHOGA) model is a cutting-edge simulation tool developed to optimize the integration of hydrogen and gas networks in the context of sustainable energy planning, particularly for the agricultural sector. This model provides a holistic framework for analyzing the production, storage, distribution, and utilization of hydrogen within gas networks, focusing on the synergy between RESs and traditional gas systems. By employing advanced optimization algorithms, iHOGA evaluates the technical and economic feasibility of various hydrogen production methods, such as electrolysis from renewable sources, and their integration into existing gas infrastructure (Hoarcă et al., 2023).

Technically, iHOGA operates by simulating the performance of hybrid energy systems under various configurations and operational strategies. It uses genetic algorithms, which are adaptive heuristic search algorithms based on the principles of natural selection and genetics, to explore a vast solution space efficiently. This approach allows iHOGA to identify optimal or near-optimal solutions for complex, non-linear, and multi-modal optimization problems that are typical in energy systems planning. The model considers various technical and economic parameters, including capital costs, maintenance costs, fuel prices, and system reliability, and can incorporate constraints such as emission limits and renewable energy targets (Hoarcă et al., 2023; Hussain et al., 2024). iHOGA's capabilities include:

- System component modeling: Detailed models for various system components (e.g., PV panels, wind turbines, batteries, diesel generators) allow for accurate simulations of their performance and interactions.
- Temporal resolution: The model supports high temporal resolution, which is essential for capturing the variability and intermittency of RESs, thus providing more reliable optimization results.
- User-friendly interface: Despite its advanced capabilities, iHOGA is designed with a user-friendly interface that facilitates its application by researchers, engineers, and policymakers, enhancing its utility in practical sustainable energy planning contexts.

8.5.12 LEAP

The Long-Range Energy Alternatives Planning (LEAP) model is a widely used, user-friendly software tool designed for integrated energy planning and policy analysis. It enables researchers, policymakers, and energy planners to assess the economic, environmental, and social impacts of different energy systems and technologies over long time horizons. LEAP employs a bottom-up approach, allowing users to create detailed energy demand and supply scenarios based on various assumptions about population growth, economic development, and technological advancements (Hu et al., 2019). This flexibility makes it suitable for modeling a wide range of energy sectors, including electricity, transportation, and heating, as well as evaluating the effects of energy efficiency measures and renewable energy integration. Its accessibility and adaptability make LEAP an invaluable tool for supporting sustainable energy planning at local, national, and global levels, fostering the transition to low-carbon energy systems and sustainable development (Sahabuddin & Khan, 2023).

8.5.13 MARKAL

The Market Allocation (MARKAL) model is a prominent optimization tool used for long-term energy planning and policy analysis, focusing on the interaction between energy supply, demand, and technology development. MARKAL employs a linear programming framework to simulate the energy system as a whole, allowing users to evaluate the cost-effective deployment of various energy technologies, including renewable sources, fossil fuels, and nuclear energy (Krzemień, 2013). By representing the entire energy supply chain, from resource extraction to end-use applications, MARKAL facilitates the analysis of complex energy systems under different scenarios, taking into account factors such as technological change, policy measures, and environmental constraints. MARKAL supports informed decision-making in sustainable energy planning, helping to identify pathways for achieving energy security, economic development, and environmental sustainability in a rapidly changing energy landscape (Friedman & Klein, 2010; Krzemień, 2013).

8.5.14 TIMES

The TIMES (The Integrated MARKAL-EFOM System) model is a powerful optimization tool designed for long-term energy planning and policy analysis, focusing on the technological and economic dynamics of energy systems. It combines features from both the MARKAL and Energy Flow Optimization Model (EFOM) frameworks to provide a comprehensive representation of energy supply and demand, enabling users to assess the impacts of various energy technologies and policies on greenhouse gas emissions, energy security, and economic development. TIMES employs a bottom-up approach, utilizing linear programming to optimize the allocation of resources across multiple sectors, including electricity, transportation, heating, and industrial processes, while considering technological advancements and market constraints (Perissi et al., 2021).

With the TIMES model, stakeholders are able to explore the implications of various policy measures, such as carbon pricing, renewable energy mandates, and energy efficiency initiatives, because it can model different energy pathways and scenarios. TIMES facilitates informed decision-making in sustainable energy planning, helping to identify optimal strategies for achieving long-term sustainability goals. Additionally, its ability to incorporate both economic and environmental factors makes it a valuable tool for evaluating the trade-offs associated with different energy strategies, ultimately supporting the transition to low-carbon energy systems that align with global climate objectives (Amorim et al., 2014).

8.5.15 TRNSYS

Transient System Simulation Tool (TRNSYS) is a versatile simulation software widely used for analyzing and designing energy systems, particularly in the context of sustainable energy planning. It provides a modular environment that allows users to create detailed models of complex energy systems, including buildings, renewable energy technologies, heating and cooling systems, and energy storage solutions. TRNSYS uses a time-dependent simulation approach to evaluate the dynamic behavior of energy systems, enabling the assessment of energy performance under varying operational conditions and climatic scenarios. This capability is crucial for understanding the interactions between different components of the energy system and optimizing their performance for energy efficiency and sustainability (Dezhdar et al., 2023).

In addition to its extensive library of pre-built components, TRNSYS can be easily integrated with other modeling tools, making it highly adaptable to a variety of sustainable energy planning applications. Users can model diverse systems, from solar thermal and photovoltaic systems to combined heat and power (CHP) and geothermal systems, allowing for comprehensive assessments of renewable energy integration and efficiency measures (Bonati et al., 2019; Villa-Arrieta & Sumper, 2018).

8.5.16 EnergyPro

EnergyPro is a comprehensive simulation and modeling software designed for the analysis and optimization of energy systems, particularly in the context of renewable energy integration and sustainable energy planning. The software offers a user-friendly interface that allows users to model various energy technologies, including photovoltaic (PV) systems, wind turbines, combined heat and power (CHP) plants, and energy storage systems. EnergyPro provides detailed analysis capabilities for assessing energy production, economic feasibility, and environmental impacts, enabling stakeholders to evaluate the performance of energy projects across multiple scenarios and configurations (Østergaard et al., 2023).

EnergyPro offers detailed lifecycle cost assessments and financial analysis, which take into account factors such as capital investment, operational costs, and revenue streams. This enables users to make informed decisions about the economic viability of renewable energy projects and to optimize their design and operation for maximum efficiency and sustainability. In addition to supporting the transition to a sustainable energy future, EnergyPro also provides insights into cost-effective strategies to reduce greenhouse gas emissions (Østergaard et al., 2022a) by facilitating the integration of RESs into existing energy systems and providing insights into cost-effective strategies for reducing greenhouse gas emissions.

8.5.17 DER-CAM

Distributed Energy Resources Customer Adoption Model (DER-CAM) is a sophisticated optimization tool developed to assess the economic viability and operational performance of DERs in various applications, including commercial and industrial settings. The model employs mixed integer linear programming (MILP) to optimize the sizing, dispatch, and operation of DERs such as solar photovoltaics, wind turbines, combined heat and power (CHP) systems, batteries, and demand response strategies. DER-CAM takes into account the unique characteristics of each facility, including energy demand profiles, utility rates, and local incentives, allowing for tailored solutions that maximize economic benefits and minimize environmental impacts (Jung & Villaran, 2017).

By providing insights into the optimal integration of DERs within the larger energy system, DER-CAM supports informed decision-making in sustainable energy planning and investment. The model's focus on the economic aspects of DER adoption makes it a valuable tool for businesses and policymakers seeking to enhance energy efficiency, reduce energy costs, and advance the transition to a more sustainable and resilient energy landscape (Mohan, 2016).

8.5.18 EnergyPLAN

EnergyPLAN is a sophisticated energy system analysis tool designed for modeling and optimizing the operation of national and regional energy systems with a focus on integrating RESs and enhancing energy efficiency (Østergaard et al., 2022b). The model utilizes a deterministic approach to simulate the dynamics of energy supply and demand, enabling users to assess the impacts of various energy policies, technology options, and market conditions on overall system performance. EnergyPLAN allows for the modeling of diverse energy sectors, including electricity, heating, transportation, and industrial processes, providing a comprehensive view of how different components interact within the energy system.

This model incorporates advanced features for simulating the integration of technologies like wind, solar, biomass, and energy storage, as well as the impacts of demand-side management and energy efficiency measures. By providing insights into optimal energy system configurations and operational strategies, EnergyPLAN supports informed decision-making in sustainable energy planning, facilitating the transition to resilient and low-carbon energy systems that align with climate and energy policy objectives (Eveloy & Ahmed, 2022; Groppi et al., 2022).

8.5.18.1 OSeMOSYS

The Open-Source Energy Modeling System (OSeMOSYS) is an advanced, linear programming-based energy systems optimization model tailored for long-term energy planning and policy analysis. It is designed to be both accessible and flexible, addressing the need for comprehensive energy modeling without the high costs associated with proprietary tools. This open-source framework allows for extensive customization and collaboration, making it an ideal tool for academic research and policy development (Groppi et al., 2023; Slimani et al., 2024).

Technically, OSeMOSYS models the entire energy system, encompassing resource extraction, energy conversion, transportation, and end-use demand. The core of OSeMOSYS consists of a linear programming problem that minimizes the total system cost while meeting energy demand and adhering to various constraints. The model structure includes parameters and variables such as technology costs, capacities, efficiencies, fuel availability, and emissions factors. For implementation, OSeMOSYS relies on a straightforward modeling language, making it accessible to users with various levels of expertise. The model code and documentation are freely available, promoting transparency and collaboration within the energy modeling community.Key features of OSeMOSYS include (Gebremeskel et al., 2023; Tan et al., 2023):

- Data inputs and structure: OSeMOSYS utilizes a wide array of input data, including energy demand projections, technology cost trajectories, resource availability, and policy targets. The model is structured in a way that facilitates the representation of different energy sectors (e.g., electricity, heating, transportation) and their interactions.
- Time resolution: The model can be configured to operate with different temporal resolutions, from annual to hourly time steps, allowing for detailed analysis of temporal variations in energy supply and demand. This feature is particularly important for integrating variable RESs.
- Geographical flexibility: OSeMOSYS can be applied to various geographical scopes, from local and national to regional and global scales. This flexibility allows for tailored analysis based on specific spatial contexts and resource distributions.

- Model outputs: The outputs from OSeMOSYS include detailed projections of energy supply and demand, technology adoption, investment requirements, operating costs, fuel use, and emissions. These outputs provide valuable insights for policymakers and stakeholders to develop robust energy strategies.

8.5.19 HOMER

Hybrid Optimization of Multiple Energy Resources (HOMER) is a flexible mathematical modeling software used to solve the problems related to the design and management of micro-grids and distributed power systems using RE and traditional power sources. The model is based on a probabilistic analysis to help the user to evaluate the economic viability as well as efficiency of different energy layouts including solar PV, wind, diesel, battery, and demand response. HOMER helps a user to model several different scenarios and to determine which configurations of energy resources yield the lowest life cycle cost while satisfying certain reliability and performance requirements (Kamal et al., 2022).

HOMER stands out for its capacity to analyze both grid-connected and off-grid systems and the potential balance that exists between investment and operational costs, emissions, and energy autonomy. Thus, sensitivity analysis and scenario exploration features of HOMER enable stakeholders to design the most effective and efficient energy systems that would harmoniously fit environmental concerns and economic priorities. Intuitively driven and equipped with superior analytical features, the software is beneficial to engineers and planners who are involved in the improvement of energy supply and distribution incorporation for sustainable energy solutions in applications such as micro-grids, industrial processes, and urban layouts (Bahramara et al., 2016).

8.6 Discussion and Classification of Modeling Tools

In the previous section, some examples of modeling tools in optimization were examined. The purpose of this review was to get familiar with the types of tools, and a brief definition and application of each of them was presented. Each of the reviewed tools belongs to a specific category as shown in Table 8.1.

- **Models of integrated energy systems**: The integrated energy system models are holistic modeling frameworks developed to analyze and optimize the whole energy system, which encompasses various sectors such as electricity, transportation, industry, and heating. They capture interactions and interdependencies between different energy carriers and technologies that enable holistic energy planning and policy analysis.

- **Building energy simulation tools**: Building energy simulation tools are specialized software employed for modeling purposes with a view to optimizing energy use in buildings. It operates on different parameters like heating, cooling, lighting, ventilation, or anything relevant to improve energy efficiency, reduce costs, and/or minimize environmental impacts.
- **DERs models**: The models for DERs are focusing on such decentralized energy resources optimization and management as solar panels, wind turbines, battery storage systems, and hydrogen technologies. They do this by fostering greater local resilience to interruptions in energy supply, reducing energy transmission losses, and integrating RESs at the local level.
- **Renewable energy planning tools**: Renewable energy planning tools are specialized models intended for planning, analysis, and optimization in integrating RESs into the existing energy system. These tools evaluate the potential, feasibility, and impacts of deploying various renewable technologies to achieve sustainability goals.
- **Models of urban energy**: The urban energy models should work out the functions of the energy systems in cities, considering residential, commercial, transportation, and industrial availing of energy. In this respect, these models deal with aspects related to peculiarities of urban energy planning that concern high energy demand density, infrastructure constraints, and the requirement for sustainable development.
- **Scenario and policy analysis models**: Scenarios and policy analysis models are used to consider different future energy scenarios and various impacts of different energy policies. All such models will help policy makers better understand how policy decisions, technology, and market forces may interact and affect each other.
- **Open-source energy models**: Open-source energy models provide available and tailorable platforms for energy system optimization. These models are open-source, so they are freely available and modifiable to adapt to specific needs by researchers and practitioners, thus promoting collaboration and innovation in energy planning.

8.7 The Future of Optimization in Energy Systems

In order to ensure the optimal operation and integration of renewable energy in energy systems, robust optimization models provide a comprehensive approach to dealing with uncertainties in energy markets. These trends and innovations offer a promising future for optimization in energy systems. They indicate a future where energy systems are more efficient, and sustainable, ultimately contributing to a cleaner and greener planet.

Table 8.1 Classification of 20 reviewed tools

Category	Description	Included models
Integrated energy system models	Comprehensive frameworks for entire energy systems across multiple sectors	MESSAGE, MARKAL, TIMES, EnergyPLAN, IEPM, Balmorel
Building energy simulation tools	Specialized tools for modeling and optimizing energy use in buildings	EnergyPlus, EnergyPro, TRNSYS
DER models	Optimization and management of decentralized energy resources	HOMER, DER-CAM, HYDROGEMS, H2RES, Hybrid2, iHOGA
Renewable energy planning tools	Tools focused on planning and optimizing the integration of RESs	RAPSIM, DEECO
Urban energy models	Models for optimizing energy systems within urban environments	URBS
Scenario and policy analysis models	Tools for exploring future energy scenarios and evaluating policy impacts	LEAP
Open-source energy models	Accessible and customizable platforms for energy system optimization	OSeMOSYS

- Advanced AI and ML: Both AI and ML are themselves evolving technologies, thereby increasing the sophistication level of optimization in energy systems. This also includes enhanced predictive analytics of energy demand, efficient integration of RESs, and advanced grid management.
- Decentralized energy systems: At the edge of the industry, decentralized energy systems, like micro-grids and DERs, are going to allow more localized production and consumption of energy ever so efficiently. These systems can be optimized using AI to balance supply and demand in real time.
- Real-time decision-making: AI and ML provide real-time analytics to energy systems to respond promptly according to demand or supply. This is valuable in smart grid management and in the optimization of energy distribution.
- Economy vs. Environment: Multi-objective optimization is able to balance the economy with environmental friendliness of regional integrated energy systems under uncertainty of renewable energy and penetration of electric vehicles (EVs).
- Balance between variability and antifragility: Steady optimization can be performed which may balance the variability-antifragility for renewable energy system design. Thus, it optimizes variability by favoring upside potential while protecting the downside toward a minimum acceptable performance.
- Co-optimization of energy and ancillary services: Robust co-optimization models could be developed for the joint dispatch of energy and ancillary service markets, which account for wind power uncertainties.

8.8 Conclusion

In this chapter, an overview of the current development of optimization techniques and models that are critical for enhancing the sustainable energy planning has been presented. The main ideas concerning the sustainable energy were described and the optimization was underlined as the crucial task for sustainable energy development. The challenges associated with these optimization techniques, such as balancing multiple conflicting objectives, managing uncertainties, and integrating diverse energy sources, were highlighted as key areas that require ongoing attention and innovation. Based on the techniques covered in this paper, we classified and described various optimization techniques of the classical and advanced type and how we were able apply them in solving some of the complications which are characteristic of the sustainable energy systems. This classification does not only provide an appropriate systematic manner of categorizing the tools but also the kinds of techniques to be applied depending on the energy planning situation at hand. In addition, the chapter analyzed the most vital modeling tools used in sustainable energy development. Energy systems are typically modeled, analyzed, and optimized using these tools; these are important to offer policy decision-makers sound paradigms on which to base their decision-making across different scenarios. The current trends in optimization techniques were also discussed, revealing a shift toward more integrated and hybrid approaches that combine multiple methodologies to leverage their strengths. These trends are driven by the need to address the increasingly complex challenges posed by the global energy transition. As for the prospects of optimization in sustainable energy planning further research and development of hybrid techniques as well as advances in computational performance and incorporation of such approaches as artificial intelligence and machine learning that in the long run seem to provide more flexible, computational and scalable solutions. In summary, this chapter underscores the critical importance of optimization in sustainable energy planning and highlights the ongoing evolution of optimization techniques and models. As we move toward a more sustainable energy future, the role of these tools will only become more prominent, necessitating continuous innovation and interdisciplinary collaboration.

References

Adeyemi-Kayode, T., Misra, S., Orovwode, H., & Adoghe, A. (2022). Modeling the next decade of energy sustainability: A case of a developing country. *Energies, 15*, 5083.

Alshammari, N. F., Samy, M. M., & Barakat, S. (2023). Comprehensive analysis of multi-objective optimization algorithms for sustainable hybrid electric vehicle charging systems. *Mathematics, 11*(7), 1741.

Amiri, M. M., Ameli, M. T., Aghamohammadi, M. R., Bashooki, E., Ameli, H., & Strbac, G. (2024). Day-ahead coordination for flexibility enhancement in hydrogen-based energy hubs in presence of EVs, storage, and integrated demand response. *IEEE Access*.

Amorim, F., Pina, A., Gerbelová, H., da Silva, P. P., Vasconcelos, J., & Martins, V. (2014). Electricity decarbonisation pathways for 2050 in Portugal: A TIMES (The Integrated MARKAL-EFOM System) based approach in closed versus open systems modelling. *Energy, 69*, 104–112.

Babatunde, O. M., Munda, J. L., & Hamam, Y. (2020). A comprehensive state-of-the-art survey on hybrid renewable energy system operations and planning. *IEEE Access, 8*, 75313–75346.

Bahramara, S., Moghaddam, M. P., & Haghifam, M. R. (2016). Optimal planning of hybrid renewable energy systems using HOMER: A review. *Renewable and Sustainable Energy Reviews, 62*, 609–620.

Bakhtavar, E., Prabatha, T., Karunathilake, H., Sadiq, R., & Hewage, K. (2020). Assessment of renewable energy-based strategies for net-zero energy communities: A planning model using multi-objective goal programming. *Journal of Cleaner Production, 272*, 122886.

Balaman, ŞY. (2016). Investment planning and strategic management of sustainable systems for clean power generation: An ε-constraint based multi objective modelling approach. *Journal of Cleaner Production, 137*, 1179–1190.

Bashir, S., Kanwal, S., Zeb, H., Baber, Z. B., & Majeed, A. (2022). Integrated energy planning and modeling (IEPM) for sustainable electricity generation in Pakistan: Challenges and limitations. *Energy Exploration & Exploitation, 40*(6), 1806–1836.

Bilal, M., Ali, M. K., Qazi, U., Hussain, S., Jahanzaib, M., & Wasim, A. (2022). A multifaceted evaluation of hybrid energy policies: The case of sustainable alternatives in special Economic Zones of the China Pakistan economic corridor (CPEC). *Sustainable Energy Technologies and Assessments, 52*, 101958.

Bonati, A., De Luca, G., Fabozzi, S., Massarotti, N., & Vanoli, L. (2019). The integration of exergy criterion in energy planning analysis for 100% renewable system. *Energy, 174*, 749–767.

Chiu, W. H., Lin, W. C., Chen, C. N., & Chen, N. P. (2021). Using an analytical hierarchy process to analyze the development of the green energy industry. *Energies, 14*(15), 4452.

Churkin, A., Bialek, J., Pozo, D., Sauma, E., & Korgin, N. (2021). Review of cooperative game theory applications in power system expansion planning. *Renewable and Sustainable Energy Reviews, 145*, 111056.

Cosic, A., Stadler, M., Mansoor, M., & Zellinger, M. (2021). Mixed-integer linear programming based optimization strategies for renewable energy communities. *Energy, 237*, 121559.

Dezhdar, A., Assareh, E., Agarwal, N., Keykhah, S., Aghajari, M., & Lee, M. (2023). Transient optimization of a new solar-wind multi-generation system for hydrogen production, desalination, clean electricity, heating, cooling, and energy storage using TRNSYS. *Renewable Energy, 208*, 512–537.

Doni, F., Gasperini, A., & Soares, J. T. (2020). What is the SDG13? In *SDG13—Climate Action: Combating Climate Change and its Impacts*. Concise Guides to the United Nations Sustainable Development Goals (pp. 21–30).

Elkasrawy, M. A., Makeen, P., Abdullatif, S. O., Memon, S., & Ghali, H. A. (2024). Investigating the utility of water cycle optimization and non-linear programming techniques in electric vehicle stations smart planning. *IETE Journal of Research, 70*(1), 597–609.

Ervural, B. C., Evren, R., & Delen, D. (2018). A multi-objective decision-making approach for sustainable energy investment planning. *Renewable Energy, 126*, 387–402.

European Commission. (2019). Resolution of the European Committee of the Regions—The Green Deal in Partnership with Local and Regional Authorities. In *Proceedings of the Communication from the Commission to the European Parliament, the Council, the European Economic and Social Committee and the Committee of the Regions* (p. 24). The European Green Deal. Available online: https://eur-lex.europa.eu/resource.html?uri=cellar:b828d165-1c22-11ea-8c1f-01aa75ed71a1.0002.02/DOC_1&format=PDF. Accessed on January 14 2023.

Eveloy, V., & Ahmed, W. (2022). Evaluation of low-carbon multi-energy options for the future UAE energy system. *Sustainable Energy Technologies and Assessments, 53*, 102584.

Fan, M., Su, D., Bhatt, M. W., & Mangal, A. (2022). Study on non-linear planning model of green building energy consumption under multi-objective optimization. *International Journal of System Assurance Engineering and Management, 13*(Suppl 1), 437–443.

Figueira, G., & Almada-Lobo, B. (2014). Hybrid simulation–optimization methods: A taxonomy and discussion. *Simulation Modelling Practice and Theory, 46*, 118–134.

Friedman, D., & Klein, Y. (2010). Urban/suburban sustainability planning: MARKAL as an analytical tool. *Middle States Geographer, 43*, 1–7.

Gebremeskel, D. H., Ahlgren, E. O., & Beyene, G. B. (2023). Long-term electricity supply modelling in the context of developing countries: The OSeMOSYS-LEAP soft-linking approach for Ethiopia. *Energy Strategy Reviews, 45*, 101045.

Genco, F., & Genco, G. (2021). Selection of energy matrix sources in Chile using a fuzzy logic decision approach. *Energy Systems, 12*, 411–429.

Gielen, D., Boshell, F., Saygin, D., Bazilian, M. D., Wagner, N., & Gorini, R. (2019). The role of renewable energy in the global energy transformation. *Energy Strategy Reviews, 24*, 38–50.

Groppi, D., Kannan, S. K. P., Gardumi, F., & Garcia, D. A. (2023). Optimal planning of energy and water systems of a small island with a hourly OSeMOSYS model. *Energy Conversion and Management, 276*, 116541.

Groppi, D., Nastasi, B., & Prina, M. G. (2022). The EPLANoptMAC model to plan the decarbonisation of the maritime transport sector of a small island. *Energy, 254*, 124342.

Güven, A. F., & Samy, M. M. (2022). Performance analysis of autonomous green energy system based on multi and hybrid metaheuristic optimization approaches. *Energy Conversion and Management, 269*, 116058.

Herc, L., Feijoo, F., Kodba, A., Dorotić, H., Stunjek, G., Beljan, D., Duić, N., et al. (2024). The management of an energy system in the realm of rapid energy transition and degasification as a consequence of energy crisis, examination in H2RES energy model. *Energy Conversion and Management, 315*, 118782.

Herc, L., Pfeifer, A., Feijoo, F., & Duić, N. (2021). Energy system transitions pathways with the new H2RES model: A comparison with existing planning tool. *e-Prime-Advances in Electrical Engineering, Electronics and Energy, 1*, 100024.

Hoarcă, I. C., Bizon, N., Şorlei, I. S., & Thounthong, P. (2023). Sizing design for a hybrid renewable power system using HOMER and iHOGA simulators. *Energies, 16*(4), 1926.

Höffner, D., & Glombik, S. (2024). Energy system planning and analysis software—A comprehensive meta-review with special attention to urban energy systems and district heating. *Energy*, 132542.

Hu, G., Ma, X., & Ji, J. (2019). Scenarios and policies for sustainable urban energy development based on LEAP model—A case study of a Postindustrial City: Shenzhen China. *Applied Energy, 238*, 876–886.

Huang, Q. (2024). Energy optimization analysis and case study of commercial buildings using EnergyPlus. *Renewable Energy Resources and Conservation*, 197–206.

Huppmann, D., Gidden, M., Fricko, O., Kolp, P., Orthofer, C., Pimmer, M., Krey, V., et al. (2019). The MESSAGEix Integrated Assessment Model and the ix modeling platform (ixmp): An open framework for integrated and cross-cutting analysis of energy, climate, the environment, and sustainable development. *Environmental Modelling & Software, 112*, 143–156.

Hussain, S., Sharma, S. K., & Lal, S. (2024). Feasible synergy between hybrid solar PV and wind system for energy supply of a green building in Kota (India): A case study using iHOGA. *Energy Conversion and Management, 315*, 118783.

Jayaraman, R., Colapinto, C., Liuzzi, D., & La Torre, D. (2017a). Planning sustainable development through a scenario-based stochastic goal programming model. *Operational Research, 17*, 789–805.

Jayaraman, R., Colapinto, C., La Torre, D., & Malik, T. (2017b). A weighted goal programming model for planning sustainable development applied to gulf cooperation council countries. *Applied Energy, 185*, 1931–1939.

Jdeed, M., Sharma, E., Klemenjak, C., & Elmenreich, W. (2018, June). Smart grid modeling and simulation—Comparing GridLAB-D and RAPSim via two case studies. In *2018 IEEE International Energy Conference (ENERGYCON)* (pp. 1–6). IEEE.

Jin, S. W., Li, Y. P., Huang, G. H., & Zhang, K. (2016). Inexact mixed-integer programming with interval-valued membership function for sustainable power-generation capacity planning. *Journal of Cleaner Production, 122*, 52–66.

Jung, J., & Villaran, M. (2017). Optimal planning and design of hybrid renewable energy systems for microgrids. *Renewable and Sustainable Energy Reviews, 75*, 180–191.

Kamal, M. M., Asharaf, I., & Fernandez, E. (2022). Optimal renewable integrated rural energy planning for sustainable energy development. *Sustainable Energy Technologies and Assessments, 53*, 102581.

Key 2024 Sustainability Trends Driving the Year Ahead. (2024). https://www.spglobal.com/esg/insights/featured/special-editorial/key-2024-sustainability-trends-driving-the-year-ahead

Khan, A. A., & Minai, A. F. (2024). A strategic review: The role of commercially available tools for planning, modelling, optimization, and performance measurement of photovoltaic systems. *Energy Harvesting and Systems, 11*(1), 20220157.

Kheybari, S., Rezaie, F. M., & Farazmand, H. (2020). Analytic network process: An overview of applications. *Applied Mathematics and Computation, 367*, 124780.

Klemm, C., & Wiese, F. (2022). Indicators for the optimization of sustainable urban energy systems based on energy system modeling. *Energy, Sustainability and Society, 12*(1), 3.

Kong, K. G. H., Lo, S. L. Y., Leong, W. D., Sunarso, J., & How, B. S. (2022). Cooperative game theory approach for multi-period inter-entity energy planning. *Chemical Engineering Transactions, 94*, 91–96.

Kotzur, L., Nolting, L., Hoffmann, M., Groß, T., Smolenko, A., Priesmann, J., Robinius, M., et al. (2021). A modeler's guide to handle complexity in energy systems optimization. *Advances in Applied Energy, 4*, 100063.

Krzemień, J. (2013). Application of Markal model generator in optimizing energy systems. *Journal of Sustainable Mining, 12*(2), 35–39.

Laveneziana, L., Prussi, M., & Chiaramonti, D. (2023). Critical review of energy planning models for the sustainable development at company level. *Energy Strategy Reviews, 49*, 101136.

Ma, X., Pan, Y., Zhang, M., Ma, J., & Yang, W. (2024). Impact of carbon emission trading and renewable energy development policy on the sustainability of electricity market: A stackelberg game analysis. *Energy Economics, 129*, 107199.

Manirambona, E., Talai, S. M., & Kimutai, S. K. (2022). Sustainability evaluation of power generation technologies using multi-criteria decision making: The Kenyan case. *Energy Reports, 8*, 14901–14914.

Mehmood, A., Iqbal, M., Zhang, L., Xiao, H., & Ren, J. (2022). An application of dimensional reduction approach for evaluating energy generation alternatives for energy sustainability. *Energy Reports, 8*, 605–614.

Mohan, M. (2016). *Utilizing SAFE (sustainable approach for energy) analysis: Informing alternative energy solutions at a campus scale* [Doctoral dissertation].

Østergaard, P. A., Andersen, A. N., & Sorknæs, P. (2022a). The business-economic energy system modelling tool energyPRO. *Energy, 257*, 124792.

Østergaard, P. A., Lund, H., Thellufsen, J. Z., Sorknæs, P., & Mathiesen, B. V. (2022b). Review and validation of EnergyPLAN. *Renewable and Sustainable Energy Reviews, 168*, 112724.

Østergaard, P. A., Johannsen, R. M., Duic, N., Lund, H., Mathiesen, B. V., Soares, M. I. R. T., & Ferreira, P. F. V. (2023). Sustainable energy planning and management. *International Journal of Sustainable Energy Planning and Management, 38*, 1–7.

Papapostolou, A., Karakosta, C., Kourti, K.-A., Doukas, H., & Psarras, J. (2019). Supporting Europe's energy policy towards a decarbonised energy system: A comparative assessment. *Sustainability, 11*, 4010.

Penna, C. C., Schot, J., & Steinmueller, W. E. (2023). Transformative investment: New rules for investing in sustainability transitions. *Environmental Innovation and Societal Transitions, 49*, 100782.

Perissi, I., Martelloni, G., Bardi, U., Natalini, D., Jones, A., Nikolaev, A., Solé, J., et al. (2021). Cross-validation of the MEDEAS energy-economy-environment model with the integrated

MARKAL-EFOM system (TIMES) and the long-range energy alternatives planning system (LEAP). *Sustainability, 13*(4), 1967.

Pfeifer, A., Feijoo, F., & Duić, N. (2023). Fast energy transition as a best strategy for all? The nash equilibrium of long-term energy planning strategies in coupled power markets. *Energy, 284*, 129109.

Radovanović, M., Filipović, S., & Panić, A. A. (2021). Sustainable energy transition in Central Asia: Status and challenges. *Energy, Sustainability and Society, 11*, 49.

Rezaei, A., Samadzadegan, B., Rasoulian, H., Ranjbar, S., Samareh Abolhassani, S., Sanei, A., & Eicker, U. (2021). A new modeling approach for low-carbon district energy system planning. *Energies, 14*(5), 1383.

Rizzo, G., & Savino, G. (2012). A linear programming model for the optimal assessment of sustainable energy action plans. *Proceedings of ECOS, 2012*, 398–401.

Rosso-Cerón, A. M., León-Cardona, D. F., & Kafarov, V. (2021). Soft computing tool for aiding the integration of hybrid sustainable renewable energy systems, case of Putumayo, Colombia. *Renewable Energy, 174*, 616–634.

Sahabuddin, M., & Khan, I. (2023). Analysis of demand, generation, and emission for long-term sustainable power system planning using LEAP: the case of Bangladesh. *Journal of Renewable and Sustainable Energy, 15*(3).

Scherer, L., Behrens, P., De Koning, A., Heijungs, R., Sprecher, B., & Tukker, A. (2018). Trade-offs between social and environmental Sustainable Development Goals. *Environmental Science & Policy, 90*, 65–72.

Sharmila, P., Baskaran, J., Nayanatara, C., & Maheswari, R. (2019). A hybrid technique of machine learning and data analytics for optimized distribution of renewable energy resources targeting smart energy management. *Procedia Computer Science, 165*, 278–284.

Shi, Y., Deng, Y., Wang, G., & Xu, J. (2020). Stackelberg equilibrium-based eco-economic approach for sustainable development of kitchen waste disposal with subsidy policy: A case study from China. *Energy, 196*, 117071.

Slimani, J., Kadrani, A., El Harraki, I., & Ezzahid, E. (2024). Towards a sustainable energy future: Modeling Morocco's transition to renewable power with enhanced OSeMOSYS model. *Energy Conversion and Management, 317*, 118857.

Streimikiene, D. (2020). Ranking of Baltic states on progress towards the main energy security goals of European energy union strategy. *Journal of International Students, 13*, 24–37.

Swain, R. B., & Karimu, A. (2019). Renewable electricity and sustainable development goals in the EU. *World Development, 125*, 104693.

Tan, N., Harrison, J., Howells, M., & Yeganyan, R. (2023). Techno-economic data and assumptions for long-term energy systems modelling in Viet Nam. *Data in Brief, 46*, 108836.

Tsangas, M., Zorpas, A.A., Jeguirim, M. (2022). Sustainable renewable energy policies and regulations, recent advances, and challenges. In *Renewable Energy Production and Distribution* (pp 449–465). Academic Press

Tutak, M., Brodny, J., Siwiec, D., Ulewicz, R., & Bindzár, P. (2020). Studying the level of sustainable energy development of the european union countries and their similarity based on the economic and demographic potential. *Energies, 13*, 6643.

Van Beuzekom, I., Gibescu, M., & Slootweg, J. G. (2015). A review of multi-energy system planning and optimization tools for sustainable urban development. *IEEE Eindhoven PowerTech, 2015*, 1–7.

Villa-Arrieta, M., & Sumper, A. (2018). A model for an economic evaluation of energy systems using TRNSYS. *Applied Energy, 215*, 765–777.

Wang, G., Chao, Y., Cao, Y., Jiang, T., Han, W., & Chen, Z. (2022). A comprehensive review of research works based on evolutionary game theory for sustainable energy development. *Energy Reports, 8*, 114–136.

Wiese, F., Bramstoft, R., Koduvere, H., Alonso, A. P., Balyk, O., Kirkerud, J. G., Ravn, H., et al. (2018). Balmorel open source energy system model. *Energy Strategy Reviews, 20*, 26–34.

Yazdanie, M., & Orehounig, K. (2023). A review and analysis of energy systems planning models and tools for renewable energy integration in cities. *Journal of Physics: Conference Series, 2600*(4), 042007.

Yu, S., You, L., & Zhou, S. (2023). A review of optimization modeling and solution methods in renewable energy systems. *Frontiers of Engineering Management, 10*(4), 640–671.

Zaheb, H., Obaidi, O., Mukhtar, S., Shirani, H., Ahmadi, M., & Yona, A. (2024). Comprehensive analysis and prioritization of sustainable energy resources using analytical hierarchy process. *Sustainability, 16*(11), 4873.

Chapter 9
Novel Prospect Theory-Based Approach to Appraise the Suppliers in Hydrogen Fuel Cell Vehicle Industry

Rishav Raj Singh, Divya Zindani, Vamsi Krishna, Sudip Dey, and Saikat Ranjan Maity

Abstract Hydrogen is anticipated to play a pivotal role in achieving goals for a low-carbon mobility future. However, the success of transitioning to a hydrogen economy hinges on well-designed supply chains with appropriate suppliers. This study delves into assessing the supplier hierarchy within the Indian context, considering the obstacles facing the hydrogen fuel cell supply chain. Through extensive literature review, the study identifies four main and twenty-three sub-challenges for evaluating hydrogen fuel cell suppliers. To analyze these suppliers, a novel prospect theory-based decision-making structure has been presented. Within the proposed approach, the linguistic uncertainties have been modeled using proportional picture fuzzy sets. Information is consolidated using proportional picture fuzzy weighted geometric (PPFWG) operator. Supplier 5 (S5) emerged as top-ranked through the proposed approach. Managerial implications of the proposed approach can aid the managers and policymakers to carry out informed decisions regarding the selection of suppliers.

Keywords Supplier selection · TODIM · Hydrogen fuel cell vehicle · Picture fuzzy sets · Proportional picture fuzzy weighted geometric (PPFWG) operator · Hydrogen fuel cell vehicles

R. R. Singh (✉) · S. Dey
Department of Mechanical Engineering, National Institute of Technology, Silchar, Assam 788010, India
e-mail: rishav_rs@mech.nits.ac.in

D. Zindani · V. Krishna
Department of Mechanical Engineering, Sri Sivasubramaniya Nadar (SSN) College of Engineering, Kalavakkam, Tamil Nadu 603110, India
e-mail: vamsi2010266@ssn.edu.in

S. R. Maity
Department of Mechanical Engineering, National Institute of Technology, Jamshedpur, Adityapur, Jharkhand 831014, India

I. Rahimi et al. (eds.), *Sustainable Supply Chain of Renewable Energy Networks*, Energy Systems in Electrical Engineering, https://doi.org/10.1007/978-981-96-4940-2_9

9.1 Introduction

The global energy transition will inevitably be one of society's most significant projects in the twenty-first century. Energy consumption amounted to almost 14,000 million tons of oil equivalent, with an expected 2.9% annual growth (Ali et al., 2021). Since they produce more than 84% of the world's total, oil and gas are the most important contributors to the global balance of essential energy sources. Transportation uses over 65% of the energy consumed overall, with oil making up 33% (Yartys et al., 2021). According to U.S. Energy estimates, burning coal accounted for 21%, natural gas accounted for 33%, and petroleum for 46% of all carbon emissions in 2019.

The transportation industry emits more CO_2 than the industrial sector because it utilizes more petroleum fuels. Light-duty vehicles (LDVs) include passenger cars and light-duty trucks, and they are responsible for the bulk of carbon emissions (Borlaug et al., 2020). In an effort to mitigate the impact of carbon dioxide and other detrimental gasses on the environment, governments are striving to swap out these vehicles with low-carbon exhaust systems. This is particularly important as the transportation industry is expected to decarbonize at a much slower rate than other industrial sectors (like power generation).

With its unique physical and chemical properties as well as a broad range of potential uses, hydrogen is anticipated to be essential to achieving the objectives to keep the transportation sector moving toward a low-carbon future. In addition to electric vehicles and biofuels, this is one more low-carbon vehicle mobility alternative. As a result, it has become the subject of ongoing research to aid in the global energy transition. According to an International Energy Agency (IEA) research, hydrogen is an energy carrier with enormous potential to improve the energy-intensive transportation sector by providing long- and short-term storage alternatives. Its primary advantage over fossil fuels is that it releases no harmful emissions and has a heating value three times more than petroleum oil.

The hydrogen economy, particularly in the context of hydrogen fuel cell vehicles (FCVs), faces significant supply chain challenges in production, storage, transportation, distribution, manufacturing, and component supply. Producing hydrogen, especially green hydrogen through electrolysis, remains cost-prohibitive due to high energy requirements and inefficiencies. For FCVs, hydrogen storage is problematic because its low volumetric energy density necessitates high-pressure tanks or cryogenic storage, both of which are technically complex and expensive. Transporting hydrogen adds another layer of difficulty, as existing natural gas infrastructure requires extensive and costly modifications to safely handle hydrogen. Furthermore, the distribution infrastructure for hydrogen refueling stations is underdeveloped, requiring substantial investment to build a network comparable to that of conventional fuel stations. Additionally, the manufacturing of fuel cells and the supply chain for critical components, such as catalysts (often platinum), membranes, and storage systems, are still in their nascent stages and require scaling up.

Selecting appropriate suppliers for the manufacturing and distribution of critical components, such as catalysts (often platinum), membranes, and storage systems, plays a pivotal role in the hydrogen economy, especially for hydrogen fuel cell vehicles (FCVs). The choice of suppliers directly impacts the cost, quality, and scalability of these components. High-quality catalysts are essential for efficient fuel cell reactions, and reliable suppliers can ensure the purity and performance consistency of these materials. Similarly, membranes must exhibit excellent conductivity and durability; sourcing them from reputable suppliers can enhance the longevity and efficiency of fuel cells. Additionally, establishing strong relationships with suppliers can lead to collaborative innovations and cost reductions through economies of scale.

Selecting appropriate suppliers for critical components in the hydrogen economy, particularly for hydrogen fuel cell vehicles (FCVs), presents several challenges. Ensuring suppliers can consistently provide high-quality materials like platinum catalysts and durable membranes is essential but difficult due to the technical expertise required. The high cost of these materials necessitates finding suppliers who offer competitive pricing without compromising on quality, which is challenging given the nascent state of the industry. Suppliers must also meet stringent industry standards and regulations for safety and environmental impact, adding further complexity.

A decision-making framework aids in circumventing supplier selection challenges in the hydrogen economy by providing a structured and systematic approach. It establishes clear evaluation criteria focusing on quality, cost, technical capability, reliability, innovation, compliance, and logistics. Although there exist numerous decision-making frameworks to assess the suppliers, the consideration of the psychological behavior of the experts to digest risk is still an ongoing topic of research. Considering the risk appetite of experts during the assessment of various suppliers is paramount for ensuring the alignment of decisions with organizational goals and risk tolerance levels. Therefore, the importance of supplier selection in the field of hydrogen fuel vehicles and the decision-making framework giving due consideration to the risk appetite of the experts, the present chapter proposes a prospect theory based dedicated decision-making framework to assess the various suppliers. The contribution of the presented chapter can be summarized as follows:

(1) Identification of key criteria/challenges faced by the suppliers catering to the hydrogen fuel cell vehicle industry.
(2) Proposition of a novel prospect theory-based decision-making structure to assess the suppliers based on the recognized challenges.

The chapter put in the order in a such a way that: Sect. 9.2 provides a brief literature review on the challenges being faced by suppliers. The preliminaries definitions required to build-up the decision-making framework have been discussed in Sect. 9.3. Section 9.3 also enumerates on the proposed decision-making framework. Section 9.4 is on application of the current decision-making structure to the case study. The managerial implications of the presented work have been summarized in Sect. 9.5 of the chapter. The chapter concludes with the concluding remarks on the presented work.

9.2 Literature Review

This section of the work summarizes the decision-making frameworks, criteria and sub criteria and the countries of investigation in the context of automotive sector. Table 9.1 summarizes the same. The research gaps and hence the novelty of the proposed approach has also been enumerated.

Following points can be summarized from the presented literature survey:

- The key barriers to implementing green supply chain management (GSCM) (Junaid, 2019; Lahane & Kant, 2019; Abdulrahman et al., 2014; Jayant & Azhar, 2014) in the automotive industry primarily revolve around human resources, financial constraints, organizational commitment, government support, and technological infrastructure. A significant challenge is the lack of skilled workforce and technical expertise, coupled with insufficient training and resistance to change. Financial limitations, such as high investment costs and inadequate funding, further hinder progress. Organizational barriers include a lack of top management commitment, unclear strategic goals, and poor stakeholder engagement. Governmental and policy-related challenges are also prominent, with insufficient legal frameworks and lack of supportive regulations for green practices. Additionally, technological and infrastructure barriers, such as inadequate IT systems and poor connectivity, impede the adoption of sustainable technologies. These issues are compounded by market competition and low supplier and consumer awareness of environmental practices, making it difficult to foster widespread adoption of GSCM in the industry
- The automotive industry employs several prominent approaches to analyze barriers to assess the suppliers within their supply chain. These include Analytic Hierarchy Process (AHP), (Anis et al., 2022; Narayana et al., 2020; Prakash et al., 2018) and Fuzzy AHP (Lahane & Kant, 2019), which are used to prioritize factors by incorporating expert judgments, with the latter adding fuzzy logic for handling uncertainty. Interpretive Structural Modeling (ISM) (Gahlot et al., 2023; Govindan et al., 2016; Jayant & Azhar, 2014; Mathiyazhagan et al., 2013) has been widely utilized for structuring complex relationships among barriers, often paired with MICMAC Analysis (Jayant & Azhar, 2014; Luthra et al., 2011) for studying influence and dependence among factors. Other prominently employed approaches are Decision-Making Trial and Evaluation Laboratory (DEMATEL) (Chand & Tarei, 2021), SWARA (Kumar et al., 2021), Fuzzy TOPSIS (Narayana et al., 2020), and WASPAS (Kumar et al., 2021). The Delphi Method (Kaviani et al., 2020), and BWM (Kaviani et al., 2020) have been employed for consensus-building and decision-making, often focusing on identifying the most critical barriers.

Several research gaps remain in the context of decision-making approaches for green supply chain management (GSCM), particularly when considering experts' risk appetite and the use of advanced fuzzy logic techniques.

- Current decision-making approaches often overlook the influence of experts' risk appetites when evaluating alternatives. There is a need for methodologies that

Table 9.1 Previous research work

Researcher	Technique	Field/country	Criteria	
			Main criteria	Sub criteria
Gahlot et al. (2023)	ISM Model	Automotive Industry/ India	• Absence of human resource, • Shortage of technical expertise, • Government support for implementing ecologically friendly policies is lacking, • Market competition, etc	
Anis et al. (2022)	AHP	Automotive Industry/ Malaysia	• Strategic barrier • Cultural barrier • Technological barrier • Individual barrier • Organizational barrier	• Absence of commitment and support from upper management, • Low index of consumer satisfaction lack of knowledge regarding SCM viewpoints on short-term decision-making, etc • Unwillingness to put supply chain procedures into practice, etc • Absence of information technology and inadequate ICT framework • Inadequate training for staff members and vendors, • Opposition to change, absence of drive and staff engagement, society's lack of awareness of social customs, etc • Absence of financial benefit, • Absence of a measurement system, etc

(continued)

Table 9.1 (continued)

Researcher	Technique	Field/country	Criteria	
			Main criteria	Sub criteria
Chand and Tarei (2021)	Grey-DEMATEL	Automobile Industry/ India	• Contingent barriers • Intra-firm barriers • Inter-firm barriers	• Unbalanced power relations, • Participant oversight, • Low reliance, etc • Passive strategy toward sustainability, • Internal cross-functional alignment, etc • Diverse sustainability standards, • Lack of faith, etc
Kumar et al. (2021)	Weighted Aggregated Sum Product Assessment methods (WASPAS), Modified Stepwise Weight Assessment Ratio Analysis (SWARA),	Indian auto component manufacturing company/ India	• Strategic barriers • Organizational barriers • Technological and infrastructure barriers • Legal barriers • Social-cultural barriers	• Deficiency of commitment from upper management, • Lack of stakeholder involvement and engagement, etc • High investment levels, • Scarcity of workforce with adequate skillset, etc • Technical standards and reference architecture are lacking, • Lack of connectivity and internet infrastructure, etc • Ineffective regulations regarding labor and employment, etc • Inadequate job opportunities for low-skilled workers, etc

(continued)

Table 9.1 (continued)

Researcher	Technique	Field/country	Criteria	
			Main criteria	Sub criteria
Urbinati et al. (2021)	SPSS, Spearman correlation test	Automotive industry/ Italy	• Product and process related • Economic • Organizational • Supply chain/customer management	• Low quality of returned goods; • High complexity of products, etc • High costs of investment and the duration of the break-even period • Little willingness to take on risk in management, etc • Rapid shift in market demands, etc
Kaviani et al. (2020)	Delphi method and BWM method	Automotive industry/Iran	• Finance-related barriers • Governance and supply chain process barriers • Knowledge-related barriers • Competitors- and market-related barriers • Management-related barriers • Policy-related barriers • Technology and infrastructure barriers	• A lack of financing for human resource training, • In the absence of initial money, etc • Problems with supply chain participants, • Forecasting and planning are limited, etc • Scarcity of information regarding RL practice, • Inadequate knowledge of RL channels, etc • Issues with underdeveloped recovery markets Customers' lack of due to the recovered product's low quality, etc • Top management engagement is low, etc • Absence of supportive laws, • Absence of explicit return and waste management guidelines, etc • Absence of newest technologies, • Insufficient of industrial Infrastructure, • Limitations on technology and obstacles to research and development pertaining to RL techniques, etc

(continued)

Table 9.1 (continued)

Researcher	Technique	Field/country	Criteria	
			Main criteria	Sub criteria
Narayana et al. (2020)	AHP and fuzzy TOPSIS	The lean and green supply chain in the medium scale automobile industrial sectors/India	• Structure of the organization, • Design of product and process, • Customer satisfaction, • Logistics, • Supplier relationships, • Environmental performance, • Capacity of the organization, • Government support, • Quality, etc	
Junaid et al. (2019)	Neutrosophic AHP and TOPSIS framework	Automotive industry/ Pakistan		• Government limitation, • Economic and environmental challenges, • Supplier insolvency, • Supplier delivery issues, • Gap in supplier communication, etc

(continued)

Table 9.1 (continued)

Researcher	Technique	Field/country	Criteria	
			Main criteria	Sub criteria
Afroz et al. (2019)	Problem Conflict Index	Automotive industry/ Malaysia	• Competitiveness and unpredictability in the market, • Not putting green practices into practice, • Insufficient knowledge, • Absence of corporate social responsibility, • Absence of globalization • Absence of technical support from the state, • Insufficient government investment, etc	

(continued)

Table 9.1 (continued)

Researcher	Technique	Field/country	Criteria	
			Main criteria	Sub criteria
Lahane and Kant (2019)	Fuzzy AHP	Supply chain management implementation in indian automobile industry/ India	• Organizational barriers • Policy and regulations barriers • Economic barriers • Technological barriers • Knowledge and support barriers	• Middle-level and lower-level managers' lack of involvement and support in the advertising of environmentally friendly goods, • Inadequate training and development initiatives for HR and SC members insufficient managerial dedication and strategy for implementing CSCM, etc • Absence of tax laws that are more favorable for advancing circular models, etc • Insufficient industry incentives to promote eco-friendly practices significant initial outlay for CSCM installation, etc • Inadequate adoption or demand for environmentally friendly technologies, • Lack of technology transfer, etc • Absence of consumer involvement in and understanding of CSC operations, • Absence of involvement and support from stakeholders, Need of skills by employees in CSCM, etc

(continued)

Table 9.1 (continued)

Researcher	Technique	Field/country	Criteria	
			Main criteria	Sub criteria
Prakash et al. (2018)	Fuzzy AHP, chang's extent analysis technique	Automobile Industry/ India	• Supply risk • Product risk • Financial risk • Demand risk	• Bad quality of raw materials, • Incorporation in business relations with suppliers, etc • Product outline risk, • Absence of skilled operators, etc • Cash flow issues, • High inventory price, etc • Transfer demand across time, • Bring demand across product, etc
Govindan et al. (2016)	Interpretive structural modeling (ISM) and analytic network process (ANP)	Automotive parts remanufacturing toward cleaner production/ India	• Business barrier • Production barrier • Technical barrier • Stakeholder barrier	• Lack of business model, • Lack of incentives, • Demand fluctuations, etc • Improper product design, • Lack of data, etc • Less expertise, • Scarcity of technologies, etc • Restrictive policies and regulations, • Lack of consumer acceptance, etc

(continued)

Table 9.1 (continued)

Researcher	Technique	Field/country	Criteria	
			Main criteria	Sub criteria
Abdulrahman et al. (2014)	Descriptive statics, Alpha coefficients, Item-total correlation, Factor analysis	Paper and paper-based products, steel/construction, plastics, automotive, electrical and electronic, and textiles. /China	• Management barriers • Financial barriers • Policy barriers • Infrastructure barriers	• Reverse logistics' importance in comparison to other challenges competitive difficulties, • Company policies, etc • Financial resources, • funds, and restrictions for handling, • Storage, etc • Legal issues/problems with supportive policies, etc • System deficiencies/EDI standards/ Underdevelopment of recycling technology, • Coordination and member support/ collaboration/reluctance to give advice, etc
Jayant et al. (2014)	MICMAC analysis and ISM approach	Auto parts manufacturing companies/India		• Cost Implication, • Insufficiency of IT applications, • Absence of official backing for GSCM adoption, • Ignorance of environmentally friendly practices, • Absence of technical knowledge, • Absence of organizational initiatives to recycle and reuse, etc

(continued)

Table 9.1 (continued)

Researcher	Technique	Field/country	Criteria	
			Main criteria	Sub criteria
Mathiyazhagan et al. (2013)	ISM approach	Auto parts manufacturing companies/India		• Issues with suppliers' continued environmental consciousness, • Insufficient government backing for the implementation of eco-friendly measures a fear of failing • Insufficient human resources • Absence of technical knowledge, • Ignorance of the environment, etc
Luthra et al. (2011)	MICMAC analysis	Automobile Industry/India		• Absence of online implementation, • Refusal to technology progress adoption, • Absence of support from the organization, • Poor human resource quality, uncertainty and competition in the market, etc

explicitly incorporate risk preferences to better align decision outcomes with expert judgments and real-world uncertainties.

- Traditional fuzzy sets using decimal values for membership, non-membership, and indecision degrees may not accurately capture human judgment. Research is needed to explore the application of proportional picture fuzzy sets (PPFSs), which allows experts to express their evaluations in relative proportions, potentially leading to more accurate and intuitive decision-making models.
- GSCM barriers have been studied extensively, there is a lack of focused research on the specific challenges related to selection of suitable supplier in the context of hydrogen fuel cell automotive industry. Identifying and addressing these barriers is crucial as the industry moves toward sustainable energy solutions.

The novelty of this work lies in three key areas:

- By developing models that integrate experts' risk preferences, this work will offer a more nuanced approach to evaluating alternatives in GSCM. Unlike traditional methods that assume a uniform risk profile, the proposed approach acknowledges the variability in risk tolerance among decision-makers, leading to decisions that better reflect the complexities and uncertainties of real-world scenarios.
- Introducing PPFS into GSCM decision-making processes represents a significant advancement in modeling expert judgments. The PPFS framework allows for the expression of membership, non-membership, and indecision degrees in relative proportions, making it easier for experts to articulate their evaluations accurately. This approach enhances the precision and reliability of the decision-making process, particularly in situations where judgments are complex and nuanced.
- The work's emphasis on identifying and addressing barriers specific to hydrogen fuel cell vehicles fills a critical gap in current GSCM research. By concentrating on this emerging area, the research contributes to the broader sustainability agenda, providing actionable insights for the automotive industry as it transitions to cleaner energy solutions.

9.3 Preliminaries

This section of the chapter discusses on the basic definitions required to build-up the proposed decision-support system. Section 9.3.1 briefs about the definitions on proportional picture fuzzy numbers (PPFNs), Sect. 9.3.2 enumerates algebraic operations on PPFNs and proportional picture fuzzy weighted geometric (PPFWG) operator. Definition on distance measures have been discussed in Sect. 9.3.3. The definitions associated with the TODIM approach elaborated in Sect. 9.3.4 and the current approach elaborated in Sect. 9.3.5.

9.3.1 Proportional Picture Fuzzy Numbers

Definition 1: Consider the picture fuzzy set $\widetilde{A} = \{x;\ \mu_{\tilde{A}}(x), \pi(x), \vartheta_{\tilde{A}}(x)|x \in X\}$. Let the expert opinion in values of $\mu(x), \vartheta(x) and \pi(x)$ as in Eqs. (9.1–9.3):

$$\mu(x) = k_1\pi(x) \tag{9.1}$$

and

$$\vartheta(x) = k_2\pi(x) \text{ satisfying } \pi(x) + k_1\pi(x) + k_2\pi(x) \le 1 \tag{9.2}$$

and

$$\pi(x) \le \frac{1}{1 + k_1 + k_2} \tag{9.3}$$

$\mu(x), \pi(x)$ and $\vartheta(x)$ are picture fuzzy numbers individually named as membership degree, indecision degree and non-membership degree. k_1, k_2 are proportional picture fuzzy numbers.

The refusal degree $r(x)$ is given by Eqs. (9.4–9.5)

$$r(x) = 1 - (\pi(x) + k_1\pi(x) + k_2\pi(x)) \tag{9.4}$$

and

$$\pi(x) = \frac{1 - r(x)}{1 + k_1 + k_2} \tag{9.5}$$

Then, each element in the set $\widetilde{A}$, defined as

$$\widetilde{A} = \left\{x;\ k_1\frac{1 - r(x)}{1 + k_1 + k_2}, \frac{1 - r(x)}{1 + k_1 + k_2}, k_2\frac{1 - r(x)}{1 + k_1 + k_2}|x \in X\right\} \tag{9.6}$$

If the $r(x) = 0$, then Eq. (9.6) becomes

$$\widetilde{A} = \left\{x;\ k_1\frac{1}{1 + k_1 + k_2}, \frac{1}{1 + k_1 + k_2}, k_2\frac{1}{1 + k_1 + k_2}|x \in X\right\} \tag{9.7}$$

Thus, the PPFs defined by Eq. (9.8):

$$\widetilde{A} = \{x;\ k_{\pi 1}, k_{\pi 2}|x \in X\} \tag{9.8}$$

9.3.2 Algebraic Operations

Definition 2: Suppose $\widetilde{A} = (\mu_{\widetilde{A}}, \pi_{\widetilde{A}}, \vartheta_{\widetilde{A}}) and \tilde{B} = (\mu_{\tilde{B}}, \pi_{\tilde{B}}, \vartheta_{\tilde{B}})$ are two PFNs. Few mathematical operations are elaborated in Eqs. (9.9) and (9.10), as below.

$$\widetilde{A} + \tilde{B} = \left(\begin{array}{l} x, k_{A1}\dfrac{1 - r_A(x)}{1 + k_{A1} + k_{A2}} + k_{B1}\dfrac{1 - r_B(x)}{1 + k_{B1} + k_{B2}} - k_{A1}\dfrac{1 - r_A(x)}{1 + k_{A1} + k_{A2}} \\ \times k_{B1}\dfrac{1 - r_B(x)}{1 + k_{B1} + k_{B2}}, \dfrac{1 - r_A(x)}{1 + k_{A1} + k_{A2}} \times \dfrac{1 - r_B(x)}{1 + k_{B1} + k_{B2}}, \\ k_{B2}\dfrac{1 - r_B(x)}{1 + k_{B1} + k_{B2}} \times k_{A2}\dfrac{1 - r_A(x)}{1 + k_{A1} + k_{A2}}, x \in X \end{array} \right) \tag{9.9}$$

$$\widetilde{A} \times \tilde{B} = \left(\begin{array}{l} x, k_{A1}\dfrac{1 - r_A(x)}{1 + k_{A1} + k_{A2}} \times k_{B1}\dfrac{1 - r_B(x)}{1 + k_{B1} + k_{B2}}, \dfrac{1 - r_A(x)}{1 + k_{A1} + k_{A2}} \\ + \dfrac{1 - r_B(x)}{1 + k_{B1} + k_{B2}} - \dfrac{1 - r_A(x)}{1 + k_{A1} + k_{A2}} \\ \times \dfrac{1 - r_B(x)}{1 + k_{B1} + k_{B2}}, k_{B2}\dfrac{1 - r_B(x)}{1 + k_{B1} + k_{B2}} + k_{A2}\dfrac{1 - r_A(x)}{1 + k_{A1} + k_{A2}} \\ - k_{B2}\dfrac{1 - r_B(x)}{1 + k_{B1} + k_{B2}} \times k_{A2}\dfrac{1 - r_A(x)}{1 + k_{A1} + k_{A2}}, x \in X \end{array} \right) \tag{9.10}$$

Equations (9.11) and (9.12) will hold true when the $r(x) = 0$

$$\widetilde{A} + \tilde{B} = \left(\begin{array}{l} k_{A1}\dfrac{1}{1 + k_{A1} + k_{A2}} + k_{B1}\dfrac{1}{1 + k_{B1} + k_{B2}} - k_{A1}\dfrac{1}{1 + k_{A1} + k_{A2}} \\ \times k_{B1}\dfrac{1}{1 + k_{B1} + k_{B2}}, \dfrac{1}{1 + k_{A1} + k_{A2}} \times \dfrac{1}{1 + k_{B1} + k_{B2}}, \\ k_{B2}\dfrac{1}{1 + k_{B1} + k_{B2}} \times k_{A2}\dfrac{1}{1 + k_{A1} + k_{A2}} \end{array} \right) \tag{9.11}$$

$$\widetilde{A} \times \tilde{B} \left(\begin{array}{l} k_{A1}\dfrac{1}{1 + k_{A1} + k_{A2}} \times k_{B1}\dfrac{1}{1 + k_{B1} + k_{B2}}, \dfrac{1}{1 + k_{A1} + k_{A2}} \\ + \dfrac{1}{1 + k_{B1} + k_{B2}} - \dfrac{1}{1 + k_{A1} + k_{A2}} \\ \times \dfrac{1}{1 + k_{B1} + k_{B2}}, k_{B2}\dfrac{1}{1 + k_{B1} + k_{B2}} + k_{A2}\dfrac{1}{1 + k_{A1} + k_{A2}} \\ - k_{B2}\dfrac{1}{1 + k_{B1} + k_{B2}} \times k_{A2}\dfrac{1}{1 + k_{A1} + k_{A2}} \end{array} \right) \tag{9.12}$$

PPFWG Operator

Definition 3 Let α_j ($j = 1, 2, \ldots., n$) be members of PPFNs. The PPFWG operator is PP^n ➔ PP such that

$$\alpha_1, \alpha_2, \ldots\ldots.\alpha_n) = \otimes_{j=1}^{n} (\alpha_j)^{w_j}$$

Here, $w = (w_1, w_2, \ldots\ldots w_n)^T$ = weight vector of $\alpha_j (j = 1, 2, \ldots n)$ and $w_j > 0$, $\sum_{j=1}^{n} w_j = 1$, then

$$PPFWGw(\alpha_1, \alpha_2, \ldots \alpha_n) = \left(\left(\prod_{j=1}^{n} (k_{ij} \frac{1}{1 + k_{1j} + k_{2j}} \right)^{w_j}, 1 - \left(\prod_{j=1}^{n} (1 - \frac{1}{1 + k_{1j} + k_{2j}} \right)^{w_j}, 1 - \left(\prod_{j=1}^{n} (1 - k_{2j} \frac{1}{1 + k_{1j} + k_{2j}} \right)^{w_j} \right)$$

9.3.3 Score Function

Definition 4: Let $X = (\mu_{xij}, \pi_{xij}, \vartheta_{xij})$ is a picture fuzzy number, the value of score function $S(X)$ explained as $S(X) = (\mu_{xij} - \vartheta_{xij})$, here $S(X) \in [-1, 1]$.

9.3.4 TODIM Method

The essential concept of the TODIM technique is the use of the prospect value function to measure the dominance or degree of dominance of each alternative relative to the others. In other words, the author may claim that this method evaluates the possibilities by assessing each alternative's partial and overall dominance over the others. It can rate the suppliers while considering the psychological qualities of Decision-makers (Gomes & Lima, 1991). The steps involved in the TODIM method's calculations are outlined below (Figueira et al., 2005; Zindani et al., 2017).

Step 1: In the first phase, the decision matrix U must be constructed, n = the number of options and m = the number of criteria.

$$U = [U_{ic}]_{n \times m} = \begin{bmatrix} x_{11} & \cdots & x_{1m} \\ \vdots & \ddots & \vdots \\ x_{n1} & \cdots & x_{nm} \end{bmatrix}$$

($i = 1, 2\ldots n$; $c = 1, 2\ldots m$).

where x_{ij} represents how well the ith alternative performed to the jth criterion.

Step 2: The choice matrix U was normalized by generating low values using Eq. (9.13) and high values recommended by the following non-beneficial criteria, which were obtained using Eq. (9.14).

$$V_{ij} = \frac{u_{ic}}{\sum_{i=1}^{n} u_{ic}} \tag{9.13}$$

$$V_{ij} = \frac{\frac{1}{u_{ic}}}{\sum_{i=1}^{n} \frac{1}{u_{ic}}} \tag{9.14}$$

where V_{ij} = normalized value of U_{ij}.

Step 3: The relative weight of the criterion for the reference criteria is derived in this phase using Eq. (9.15):

$$w_{tr} = \frac{w_t}{w_r} \tag{9.15}$$

Here, w_{tr} = relative weights, w_t = weight of reference criteria.

Step 4: By employing Eq. (9.16), the dominance level of E_i over the alternative E_j is determined.

$$\delta(E_i, E_j) = \sum_{c=1}^{m} \Phi_c(E_i, E_j) \tag{9.16}$$

where $\Phi_c(E_i, E_j) = \begin{cases} \sqrt{\frac{rw_{cr}(V_{ic}-V_{jc})}{\sum_{c=1}^{m} rw_{cr}}} & \text{if } (V_{ic} - V_{jc}) > 0 \\ 0 & \text{if } (V_{ic} - V_{jc}) = 0 \\ \frac{-1}{\ominus}\sqrt{\frac{(\sum_{c=1}^{m} rw_{cr})(V_{ic}-V_{jc})}{rw_{cr}}} & \text{if } (V_{ic} - V_{jc}) < 0 \end{cases}$

where $\Phi_c(E_i, E_j)$ dominance value of substitute E_i over E_j, rw_{cr} = relative weights, and $\ominus$ is attenuation factor.

The gains and losses of the ith option over the jth alternative are represented in the conditions mentioned above by $(V_{ic} - V_{jc}) > 0$ and $(V_{ic} - V_{jc}) < 0$, respectively, signifies the attenuation factor of the loss. In TODIM, the negative quadrant represents alternatives that are less favorable or perform worse relative to the reference criterion or point of comparison. Decision-makers are more sensitive to losses than gains, so the weight assigned to losses in the negative quadrant is higher to reflect this psychological aversion to losses.

Step 5: Eq. (9.17) calculates the degree of total dominance of option E_i (£$_i$).

$$£_i = \frac{\sum_{j=1}^{n} \delta(E_i, E_j) - \min \sum_{j=1}^{n} \delta(E_i, E_j)}{\max \sum_{j=1}^{n} \delta(E_i, E_j) - \min \sum_{j=1}^{n} \delta(E_i, E_j)} \tag{9.17}$$

where £$_i$ total dominance value.

Step 6: All alternatives have been ranked according to dominance values in the final stage and highest dominance score deemed as best option.

9.3.5 Suggested Approach

Figure 9.1 demonstrates the different phases of the proposed decision-making framework. The phases have been discussed as follows:

Phase 1: Initialization phase: First phase is to identify the criteria/sub criteria barriers for the hydrogen fuel cell. This is followed with the identification of the suppliers to be appraised.

Phase 2: Modeling of linguistic evaluations phase: Second phase of the proposed approach is collection of linguistic values from experts and convert those values into PPFS.

Phase 3: Aggregation of expert information phase: In this phase Def. 3 is used to aggregate the information received from different experts, subsequently the score values are calculated as per the definition in Sect. 9.3.3.

Phase 4: Ranking of supplier's phase: In this phase, the score value matrix is normalized and subsequently weights, and relative weights (Eq. 9.15) are calculated. Then TODIM is used to evaluate the overall dominance scores and hence rank the suppliers.

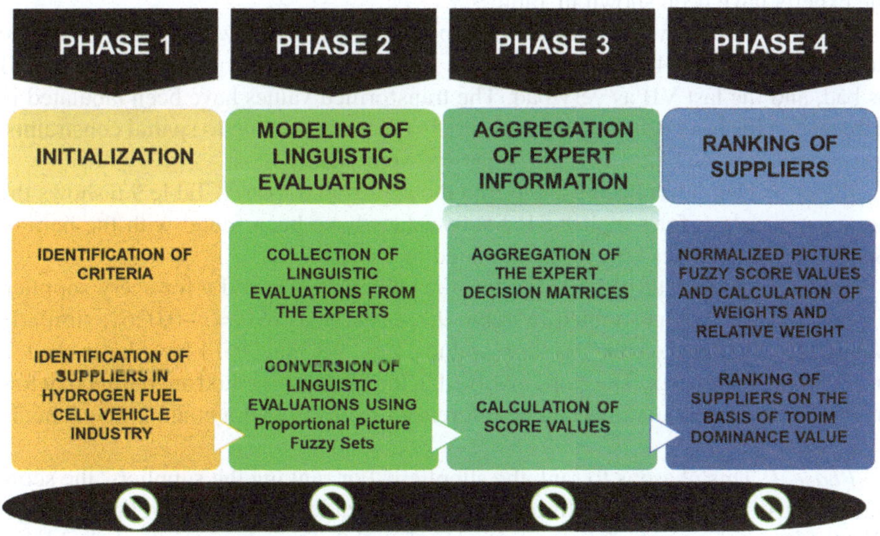

Fig. 9.1 Plan of steps involved in the current decision-making framework

9.4 Case Study

The case study relates to appraisal of suppliers for supplying critical hydrogen fuel cell component. This has been done through the application of the current approach. The identified suppliers have been evaluated based on various criteria and sub criteria which were chosen from the past case studies, experts' opinion, and pilot study through questionnaire. After choosing the criteria/sub criteria, an elaborated questionnaire has been prepared which first used for pilot study and as per the feedback it has been modified with the help of experts. These data have been used further to analyze them through the proposed decision-making approach. All the steps have been illustrated in the flow chart, followed by different definitions and formula used in the study. The steps have been followed to appraise the identified suppliers.

Phase 1: First phase of the suggested approach is to identify criteria and sub criteria. This has been obtained through the studied literature. In the present context, 23 criteria/sub criteria have been considered. These have been summarized in Table 9.2. Three experts EX1, EX2, and EX3 have been chosen to assess five suppliers S1, S2.... S5 based on the identified criteria and sub criteria, respectively.

Phase 2: This phase relates to the collection of data and conversion to linguistic variables through the usage of proportional picture fuzzy sets. Evaluations from all the experts have been shown in Table 9.3.

Table 9.3 has been transformed to proportional picture fuzzy numbers as per the depiction in Table 9.4, where VG read as very good, G as good, M as moderate, B as bad, and the last VB as very bad. The transformed values have been tabulated in Table 9.5. This has been shown for expert 1 (EX1) only owing to spatial constraints. This completes the phase 2 of the suggested approach.

Phase 3: Def. 3 has been used to find the aggregated matrix. Table 9.6 shows the aggregated values for supplier S1. Score values have been found with the help of Sect. 9.3.3, and values have been tabulated in Table 9.7.

The aggregated matrix Table 9.6 has the values of μ, π, and υ for every supplier. To find the score values tabled in Table 9.7, 0.4429–0.4690 $= -0.0261$, similarly others are calculated. After the score values, Eqs. 9.13 and 9.14 have been used to normalizing the matrix. The steps involved in the TODIM method are outlined below and the detailed calculations can be referenced from (Figueira et al., 2005; Zindani et al., 2017).

Phase 4: This phase is to rank the suppliers. For ranking the suppliers, the score value matrix has been normalized. After normalization weight and relative weight (Eq. 9.15) have been calculated and shown in Fig. 9.2. The sub criteria have the higher values of weight, signifies higher importance among the rest of the sub criteria. For example, RB5 has more importance than EB2, EB2 etc.

After the normalization of the score values, Eqs. (9.16) and (9.17) have been used to find the overall dominance values and ranking of the suppliers. The overall dominance values and ranking are tabulated in Table 9.8. This completes the completed suggested approach.

Table 9.2 Shortlisted barriers/criteria and sub criteria from the previous studies

Main barriers	Nomenclature	Sub barriers	Definition of the barriers	References
Economic barriers	EB			
	EB1	Cost of fuel cells	The overall price of the fuel cells	(Maestre et al., 2021; Wu et al., 2021)
	EB2	The production cost of hydrogen	Cost to manufacture the product	(Maestre et al., 2021)
	EB3	Purchase costs	Price to purchase the end product	(Wu et al., 2021)
	EB4	Market size	Size of the market to sell the product	(Maestre et al., 2021)
	EB5	Consumer demand	Ask for the product in the public	(Al-Amin & Doberstein, 2019; Wu et al., 2021)
Technical barriers	*TB*			
	TB1	Lack of infrastructures	Less physical facilities to provide the services	(Wu et al., 2021); (Maestre et al., 2021)
	TB2	Policies	Government support	(Al-Amin & Doberstein, 2019; Maestre et al., 2021)
	TB3	Maintenance issues	Supporting infrastructure to continuously run the system	(Al-Amin & Doberstein, 2019)
	TB4	Durability	Life of the product	(Maestre et al., 2021)
	TB5	Skilled human resources	Knowledgeable persons in the field	(Al-Amin & Doberstein, 2019; Maestre et al., 2021)
	TB6	Availability of hydrogen fuel cell	Accessibility of hydrogen fuel cell in the market	(Maestre et al., 2021)
	TB7	Storage capacity	Capacity to store the hydrogen	(Maestre et al., 2021)
Social barriers	*SB*			
	SB1	Public awareness	Knowledge to the public	(Wu et al., 2021)

(continued)

Table 9.2 (continued)

Main barriers	Nomenclature	Sub barriers	Definition of the barriers	References
	SB2	Social acceptance	Acceptance of the product socially	(Al-Amin & Doberstein, 2019; Wu et al., 2021)
	SB3	Customer paying capacity	Capacity to buy the product	(Al-Amin & Doberstein, 2019)
	SB4	Safety	Protection from the mishap	(Maestre et al., 2021; Wu et al., 2021)
Regulatory barriers	*RB*			
	RB1	Tax policies	Favorable tax polices	(Al-Amin & Doberstein, 2019; Maestre et al., 2021)
	RB2	Rules to support manufacturers	Manufacturers motivating policies	(Wu et al., 2021)
	RB3	Communication gap	How quick messages reached to the last element of the system	(Wu et al., 2021)
	RB4	Environmental rules	Nature-friendly polices	(Al-Amin & Doberstein, 2019)

As per the ranking results, supplier S5 has been found to be the most suitable supplier among the considered ones whereas supplier S3 has been found to be the least preferred supplier. Figure 9.3 shows graph of dominance values and rank.

9.4.1 *Sensitivity Analysis for the Different Values of the Attenuation Factor (Θ)*

The robustness of the proposed methodology has been checked by changing the values of authentication factor (Θ) from 10 to 0.1 and same in Fig. 9.4 has been shown, where Θ values on X- axis and ranking values on Y-axis. Supplier S5 is most preferable and S3 is least preferable among the suppliers.

Figure 9.5 shows the graph between Θ values and dominance values which have been received after changes the Θ values from 10 to 0.1.

Table 9.3 Response of all the three experts for 23 criteria/sub criteria

	EX1					EX2					EX3				
	S1	S2	S3	S4	S5	S1	S2	S3	S4	S5	S1	S2	S3	S4	S5
EB1	G	VG	M	G	M	B	M	VB	B	B	M	B	M	B	B
EB2	B	M	M	M	G	VB	B	B	VB	VB	B	VB	M	B	B
EB3	VG	M	B	M	G	M	B	M	M	B	VB	B	M	VB	B
EB4	VG	G	VG	G	VG	G	VG	VG	G	G	VG	G	VG	M	G
EB5	B	VB	B	VB	B	VB	B	B	B	VB	VB	B	VB	VB	VB
TB1	M	VB	B	VB	B	M	B	B	M	B	VB	VB	VB	B	VB
TB2	G	M	G	G	G	G	G	VG	VG	G	G	M	VG	G	G
TB3	B	M	B	B	M	B	B	M	M	M	M	B	M	M	B
TB4	M	M	B	G	VG	G	G	VG	VG	M	G	M	G	G	VG
TB5	VB	VB	VB	VB	B	VB	B	B	B	M	B	B	VB	VB	B
TB6	M	VB	M	B	B	B	VB	VB	B	B	B	VB	B	VB	B
TB7	G	VG	G	M	M	G	G	M	G	G	M	G	VG	G	G
SB1	M	VB	VB	B	VB	VB	VB	VB	B	VB	M	B	B	VB	B
SB2	VB	VB	B	VB	VB	B	M	B	M	M	VB	B	M	B	B
SB3	B	M	G	VG	VG	G	VG	VG	G	G	G	M	VG	G	G
SB4	VB	VB	VB	VB	B	VB	B	B	VB	VB	B	M	B	VB	VB
SB5	G	G	G	VG	VG	G	VG	G	VG	VG	G	VG	G	VG	VG
RB1	M	G	G	M	G	B	M	M	B	B	VB	B	VB	M	M
RB2	G	VG	M	G	G	G	VG	M	G	VG	M	G	G	VG	VG
RB3	M	B	B	M	B	M	M	B	B	B	M	G	M	G	G
RB4	G	M	M	M	G	B	G	M	G	G	B	M	M	G	B
RB5	B	M	G	M	B	G	M	M	G	G	B	G	G	M	G
RB6	M	B	B	B	M	G	VG	G	VG	VG	G	G	VG	M	M

Table 9.4 Parameter for converting decision matrix inputs

Linguistic variables	Proportional picture fuzzy numbers	
VG	8	2
G	7	3
M	5	5
B	3	7
VB	2	8

Table 9.5 Transformed values of EX1 input

	S1		S2		S3		S4		S5	
EB1	7	3	8	2	5	5	7	3	5	5
EB2	3	7	5	5	5	5	5	5	7	3
EB3	8	2	5	5	3	7	5	5	7	3
EB4	8	2	7	3	8	2	7	3	8	2
EB5	3	7	2	8	3	7	2	8	3	7
TB1	5	5	2	8	3	7	2	8	3	7
TB2	7	3	5	5	7	3	7	3	7	3
TB3	3	7	5	5	3	7	3	7	5	5
TB4	5	5	5	5	3	7	7	3	8	2
TB5	2	8	2	8	2	8	2	8	3	7
TB6	5	5	2	8	5	5	3	7	3	7
TB7	7	3	8	2	7	3	5	5	5	5
SB1	5	5	2	8	2	8	3	7	2	8
SB2	2	8	2	8	3	7	2	8	2	8
SB3	3	7	5	5	7	3	8	2	8	2
SB4	2	8	2	8	2	8	2	8	3	7
SB5	7	3	7	3	7	3	8	2	8	2
RB1	5	5	7	3	7	3	5	5	7	3
RB2	7	3	8	2	5	5	7	3	7	3
RB3	5	5	3	7	3	7	5	5	3	7
RB4	7	3	5	5	5	5	5	5	7	3
RB5	3	7	5	5	7	3	5	5	3	7
RB6	5	5	3	7	3	7	3	7	5	5

9.5 Managerial Implications

One effective method for handling the supplier selection procedure is demonstrated by the decision-making strategy that is being given. The decision-making framework that has been developed can be utilized by the mobility space authorities to evaluate different providers while taking into account different obstacles, including technological, financial, social, and regulatory restrictions. Before selecting a provider for their purchase, industry may be aware of the possible obstacles associated with various suppliers. The issue of adding any incompatible supplier to the supply chain network can be avoided with the help of the prudent decision-making made possible by the suggested decision-making system. Additionally, it will assist the companies in saving a significant amount of money and time in the event of unforeseen circumstances. By producing hydrogen fuel cell cars in large quantities and securing a stable long-term supply, their monetary value can be reduced. Because the recommended

Table 9.6 Aggregated matrix

S1			
EB1	0.4429	0.0846	0.4690
EB2	0.2515	0.0909	0.6567
EB3	0.3610	0.0834	0.5435
EB4	0.5591	0.0735	0.3674
EB5	0.2182	0.0909	0.6896
TB1	0.3298	0.0909	0.5720
TB2	0.5669	0.0825	0.3509
TB3	0.3261	0.0909	0.5809
TB4	0.5290	0.0881	0.3820
TB5	0.2095	0.0909	0.6984
TB6	0.3432	0.0909	0.5636
TB7	0.4996	0.0832	0.4167
SB1	0.3784	0.0909	0.5252
SB2	0.1972	0.0909	0.7111
SB3	0.4346	0.0909	0.4676
SB4	0.2095	0.0909	0.6984
SB5	0.5385	0.0769	0.3846
RB1	0.2978	0.0909	0.6054
RB2	0.5247	0.0846	0.3900
RB3	0.4545	0.0909	0.4545
RB4	0.3646	0.0860	0.5445
RB5	0.3103	0.0887	0.5981
RB6	0.5096	0.0871	0.4026

decision-making method is employable, managers can choose the best supplier for their supply chain network and give their companies a competitive edge.

The decision-making framework for supplier selection, while highly effective in the mobility sector, may face challenges in implementation across different industries. Each sector may prioritize distinct factors—such as technology compatibility, emissions standards, or cost-efficiency—which could necessitate tailoring the framework to fit specific needs. Additionally, scalability could become an issue, particularly in industries with complex global supply chains or limited supplier options. Adapting the framework for broader application would require balancing its flexibility and adaptability while maintaining its ability to account for the technological, financial, social, and regulatory obstacles that different sectors encounter.

Table 9.7 Score values

	S1	S2	S3	S4	S5
EB1	−0.0261	0.0377	−0.1467	−0.1798	−0.2204
EB2	−0.4052	−0.3076	−0.0866	−0.2715	−0.2234
EB3	−0.1825	−0.2204	−0.1843	−0.2422	−0.1798
EB4	0.1918	0.1480	0.1852	0.1831	0.1736
EB5	−0.4714	−0.4533	−0.4346	−0.5139	−0.4714
TB1	−0.2422	−0.5139	−0.4346	−0.4018	−0.4346
TB2	0.2161	0.0639	0.2793	0.2106	0.2439
TB3	−0.2548	−0.2204	−0.1843	−0.1843	−0.1466
TB4	0.1469	0.0229	−0.1089	0.1600	0.1581
TB5	−0.4888	−0.4533	−0.5139	−0.5139	−0.3036
TB6	−0.2204	−0.5455	−0.2715	−0.4346	−0.3636
TB7	0.0828	0.2208	0.1167	0.0818	0.0706
SB1	−0.1467	−0.4888	−0.4888	−0.4346	−0.4888
SB2	−0.5139	−0.4018	−0.2548	−0.4018	−0.4018
SB3	−0.0330	0.0395	0.2453	0.1609	0.2075
SB4	−0.4888	−0.3601	−0.4533	−0.5455	−0.4714
SB5	0.1538	0.2603	0.1285	0.3212	0.1982
RB1	−0.3076	−0.0190	−0.1291	−0.0866	0.0500
RB2	0.1347	0.2509	0.0404	0.1883	0.2793
RB3	0.0001	−0.1515	−0.2548	−0.0498	−0.2168
RB4	−0.1798	0.0639	−0.0001	0.0754	−0.0447
RB5	−0.4295	0.1146	0.1212	0.0639	−0.1321
RB6	0.1070	−0.1181	−0.1096	−0.1522	0.0395

9.6 Conclusions

In conclusion, this work aims to bridge significant research gaps in green supply chain management (GSCM) by introducing innovative approaches that enhance decision-making and address emerging challenges in the automotive industry. By incorporating experts' risk appetites into decision-making models, the research offers a more sophisticated and realistic framework that accounts for the variability in risk tolerance among stakeholders. The application of proportional picture fuzzy sets (PPFS) further refines the decision-making process, enabling more accurate and intuitive representation of expert judgments. Additionally, the focused exploration of barriers related to hydrogen fuel cell vehicles underscores the importance of addressing specific challenges as the industry moves toward sustainable energy solutions. The complete phase wise process shows the ranking in the order of S5 > S1 > S4 > S2 > S3 which means Supplier 5 (S5) is the best supplier than all the other suppliers. The differences in

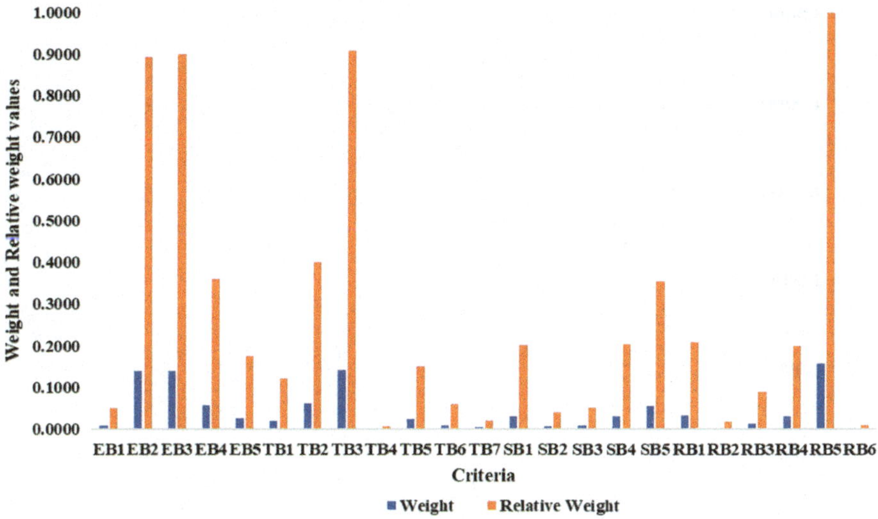

Fig. 9.2 Weight and relative weight of the criteria

Table 9.8 Dominance vales and rank of the suppliers

Suppliers	Dominance value	Rank
S5	1.0000	1
S1	0.6427	2
S4	0.5938	3
S2	0.4507	4
S3	0.0000	5

ranking results unequivocally point to the influence of the expert's risk tolerance when making a choice. The future scope of this study is broad and holds significant potential for advancing the field of green supply chain management (GSCM). One promising direction is the further development and refinement of decision-making models that integrate experts' risk appetites. This could involve exploring dynamic risk preferences, incorporating scenario-based risk assessments, and evaluating the impact of collective risk perceptions in group decision contexts. Additionally, the application of proportional picture fuzzy sets (PPFSs) can be expanded beyond GSCM to other complex decision-making areas, such as healthcare, finance, and public policy, where accurate representation of expert judgments is crucial. This expansion could lead to the creation of more versatile and intuitive decision-support tools. Moreover, addressing specific barriers related to hydrogen fuel cell vehicles offers an opportunity to advance sustainable energy solutions and improve the adoption of cleaner technologies.

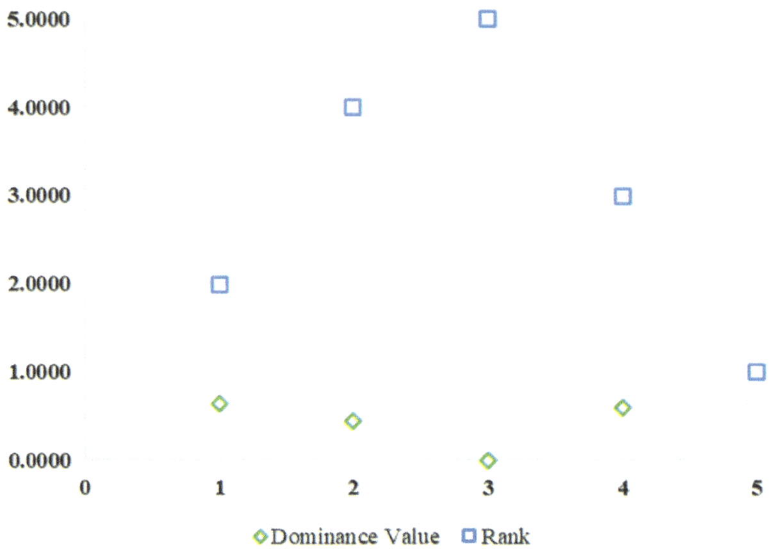

Fig. 9.3 Graph between dominance values and rank of the suppliers

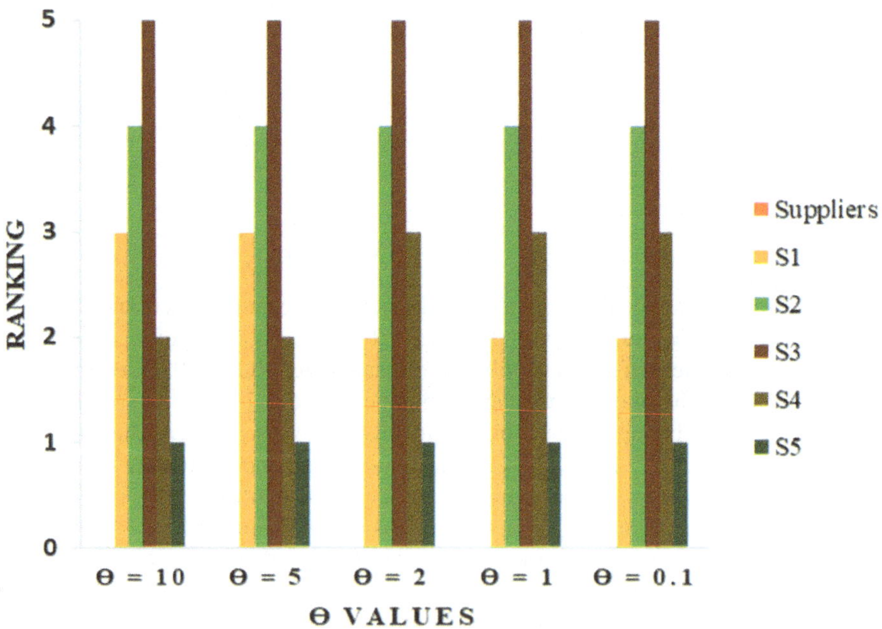

Fig. 9.4 Graph between Θ values and ranking for different Θ values

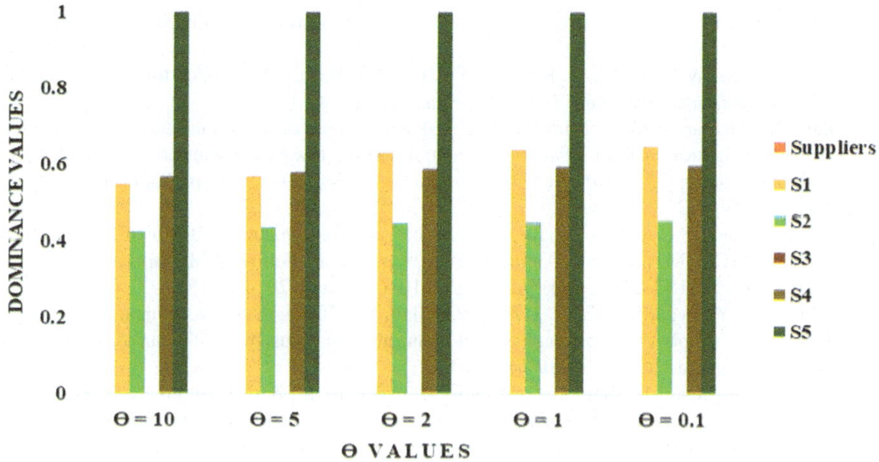

Fig. 9.5 Graph between Θ values and dominance values for different Θ values

References

Abdulrahman, M. D., Gunasekaran, A., & Subramanian, N. (2014). Critical barriers in implementing reverse logistics in the Chinese manufacturing sectors. *International Journal of Production Economics, 147*, 460–471. https://doi.org/10.1016/j.ijpe.2012.08.003

Afroz, R., Rahman, A., Muhibbullah, M., & Morshed, N. (2019). Malaysian automobile industry and green supply chain management. *International Journal of Recent Technology and Engineering, 7*(6), 158–162.

Al-Amin, A. Q., & Doberstein, B. (2019). Introduction of hydrogen fuel cell vehicles: Prospects and challenges for Malaysia's transition to a low-carbon economy. *Environmental Science and Pollution Research, 26*(30), 31062–31076. https://doi.org/10.1007/s11356-019-06128-4

Ali, S. A., Alharthi, M., Hussain, H. I., Rasul, F., Hanif, I., Haider, J., & Abbas, Q. (2021). A clean technological innovation and eco-efficiency enhancement: A multi-index assessment of sustainable economic and environmental management. *Technological Forecasting and Social Change, 166*, 120573. https://doi.org/10.1016/j.techfore.2021.120573

Anis, A., Islam, R., & Hafit, N. I. A. (2022). Prioritisation of barriers for supply chain management implementation in the Malaysian automotive industry. *Production, 32*.https://doi.org/10.1590/0103-6513.20210073

Borlaug, B., Holden, J., Wood, E., Lee, B., Fink, J., Agnew, S., & Lustbader, J. (2020). Estimating region-specific fuel economy in the United States from real-world driving cycles. *Transportation Research Part D: Transport and Environment, 86*, 102448. https://doi.org/10.1016/j.trd.2020.102448

Chand, P., & Tarei, P. K. (2021). Do the barriers of multi-tier sustainable supply chain interact? A multi-sector examination using resource-based theory and resource-dependence theory. *Journal of Purchasing and Supply Management, 27*(5), 100722. https://doi.org/10.1016/j.pursup.2021.100722

Figueira, J., Greco, S., & Ehrgott, M. (2005). *Multiple criteria decision analysis: state of the art surveys* (5th ed.). Springer.

Gahlot, N. K., Bagri, G. P., Gulati, B., Bhatia, L., & Das, S. (2023). Analysis of barriers to implement green supply chain management practices in Indian automotive industries with the help of

ISM model. *Materials Today: Proceedings, 82*, 330–339. https://doi.org/10.1016/j.matpr.2023.02.146
Gomes, L. F. A. M., & Lima, M. M. P. P. (1991). TODIMI: Basics and application to multicriteria ranking. *Foundation Computing Decision Sciences, 16*(3–4), 1–16.
Govindan, K., Shankar, K. M., & Kannan, D. (2016). Application of fuzzy analytic network process for barrier evaluation in automotive parts remanufacturing towards cleaner production–A study in an Indian scenario. *Journal of Cleaner Production, 114*, 199–213. https://doi.org/10.1016/j.jclepro.2015.06.092
Jayant, A., & Azhar, M. (2014). Analysis of the barriers for implementing green supply chain management (GSCM) practices: An interpretive structural modeling (ISM) approach. *Procedia Engineering, 97*, 2157–2166. https://doi.org/10.1016/j.proeng.2014.12.459
Junaid, M., Xue, Y., Syed, M. W., Li, J. Z., & Ziaullah, M. (2019). A neutrosophic ahp and topsis framework for supply chain risk assessment in automotive industry of Pakistan. *Sustainability, 12*(1), 154. https://doi.org/10.3390/su12010154
Kaviani, M. A., Tavana, M., Kumar, A., Michnik, J., Niknam, R., & de Campos, E. A. R. (2020). An integrated framework for evaluating the barriers to successful implementation of reverse logistics in the automotive industry. *Journal of Cleaner Production, 272*, 122714. https://doi.org/10.1016/j.jclepro.2020.122714
Kumar, V., Vrat, P., & Shankar, R. (2021). Prioritization of strategies to overcome the barriers in industry 4.0: A hybrid MCDM approach. *Opsearch*, 1–40. https://doi.org/10.1007/s12597-020-00505-1
Lahane, S., & Kant, R. (2019). Identification and prioritization of barriers to circular supply chain management implementation in Indian automobile industry: A fuzzy AHP approach. *61st 5th*, 57.
Luthra, S., Kumar, V., Kumar, S., & Haleem, A. (2011). Barriers to implement green supply chain management in automobile industry using interpretive structural modeling technique: An Indian perspective. *Journal of Industrial Engineering and Management (JIEM), 4*(2), 231–257. https://doi.org/10.3926/jiem.v4n2.p231-257
Maestre, V., Ortiz, A., & Ortiz, I. (2021). Challenges and prospects of renewable hydrogen-based strategies for full decarbonization of stationary power applications. *Renewable and Sustainable Energy Reviews, 152*, 111628.
Mathiyazhagan, K., Govindan, K., NoorulHaq, A., & Geng, Y. (2013). An ISM approach for the barrier analysis in implementing green supply chain management. *Journal of Cleaner Production, 47*, 283–297. https://doi.org/10.1016/j.jclepro.2012.10.042
Narayana, S. N., Parthiban, P., & Dhanalakshmi, R. (2020). Assessment and implementation of lean and green supply chain in medium scale automobile industries using AHP and fuzzy TOPSIS. *Journal of Scientific and Industrial Research, 79*, 720–726.
Prakash, A., Agarwal, A., & Kumar, A. (2018). Risk assessment in automobile supply chain. *Materials Today: Proceedings, 5*(2), 3571–3580. https://doi.org/10.1016/j.matpr.2017.11.606
Urbinati, A., Franzò, S., & Chiaroni, D. (2021). Enablers and barriers for circular business models: An empirical analysis in the Italian automotive industry. *Sustainable Production and Consumption, 27*, 551–566. https://doi.org/10.1016/j.spc.2021.01.022
Wu, Y., Liu, F., He, J., Wu, M., & Ke, Y. (2021). Obstacle identification, analysis and solutions of hydrogen fuel cell vehicles for application in China under the carbon neutrality target. *Energy Policy, 159*, 112643. https://doi.org/10.1016/j.enpol.2021.112643
Yartys, V. A., Lototskyy, M. V., Linkov, V., Pasupathi, S., Davids, M. W., Tolj, I., ... & Suwarno, S. (2021). Hydride4mobility: An EU HORIZON 2020 project on hydrogen powered fuel cell utility vehicles using metal hydrides in hydrogen storage and refuelling systems. *International Journal Of Hydrogen Energy, 46*(72), 35896–35909. https://doi.org/10.1016/j.ijhydene.2021.01.190

Zindani, D., Maity, S. R., Bhowmik, S., & Chakraborty, S. (2017). A material selection approach using the TODIM (TOmada de Decisao Interativa Multicriterio) method and its analysis. *International Journal of Materials Research, 108*(5), 345–354. https://doi.org/10.3139/146.111489

Chapter 10
RenewGenix: A Next-Generation Open-Source Polygeneration Simulation Platform for Sustainable Energy Systems

Ismail Baklouti

Abstract In the era of sustainable development, optimizing energy systems for efficiency and environmental impact is paramount. RenewGenix emerges as a pioneering open-source software platform designed to simulate and optimize polygeneration energy systems. With a core planned to be developed in C++ and leveraging the computational fluid dynamics capabilities of OpenFOAM, RenewGenix will aim to offer unprecedented accuracy in modeling complex renewable energy networks. This chapter introduces the conceptual framework of RenewGenix, outlines its potential for revolutionizing sustainable energy supply chains, and presents a roadmap for its development and deployment. It encapsulates the philosophy behind its open-source approach, fostering collaboration and innovation in renewable energy technologies. Through RenewGenix, we envision a future where polygeneration systems are seamlessly integrated, facilitating a global transition toward cleaner, more efficient, and sustainable energy production and management.

Keywords Polygeneration · Sustainable energy systems · Simulation platform · Open-source · Optimization

10.1 Introduction

10.1.1 Addressing Gaps and Motivating RenewGenix

The transition to sustainable energy systems necessitates innovative approaches to modeling, simulating, and optimizing complex energy networks. While existing tools and research efforts have made significant progress in addressing specific aspects of energy systems, such as home energy management (Javadi et al., 2021; Nezhad et al., 2022), energy hub operation (Javadi et al., 2019), and wind-hydrothermal scheduling (Esmaeel Nezhad et al., 2024), a crucial gap remains in providing a comprehensive and open-source platform that specifically targets polygeneration systems.

I. Baklouti (✉)
National School of Engineers of Monastir, University of Monastir, Monastir, Tunisia
e-mail: ismail.baklouti@enim.u-monastir.tn

I. Rahimi et al. (eds.), *Sustainable Supply Chain of Renewable Energy Networks*, Energy Systems in Electrical Engineering, https://doi.org/10.1007/978-981-96-4940-2_10

Polygeneration, the simultaneous production of multiple energy carriers (electricity, heat, cooling) from a single energy source, offers significant potential for enhancing energy efficiency and reducing environmental impact. However, designing and optimizing these integrated systems require a holistic approach that considers the complex interactions between diverse energy technologies. RenewGenix emerges as a response to this need, providing a dedicated platform for modeling, simulating, and optimizing polygeneration systems, bridging the gap between existing tools and the increasing complexity of sustainable energy networks.

10.1.2 The Global Energy Landscape

The world stands at a critical juncture, facing an unprecedented energy challenge. Climate change, undeniably driven by our historical reliance on fossil fuels, presents an urgent and compelling demand for a rapid transition to sustainable energy sources (Groppi et al., 2018; Herc et al., 2022). This shift is not merely a matter of environmental responsibility but an absolute necessity for the health and well-being of our planet and future generations. Resource depletion, coupled with ever-growing global energy demands, further underscores the urgency for change, pushing us to explore innovative and sustainable alternatives (Alrai et al., 2024; Saritha et al., 2021).

While renewable energy sources like solar and wind power offer promising solutions, their inherent intermittency, fluctuating with weather patterns and time of day, presents a significant hurdle in our quest for a sustainable energy future. This inherent variability poses a challenge for maintaining a stable and reliable energy grid, accustomed to the consistent output of traditional power plants. Addressing this intermittency issue requires the development of sophisticated energy storage solutions and intelligent grid management systems capable of balancing supply and demand in real-time, ensuring a seamless transition to a future powered predominantly by renewable sources.

10.1.3 Polygeneration: A Pathway to Sustainability

In the pursuit of sustainable energy solutions, polygeneration emerges as a beacon of hope, offering a compelling pathway toward a more efficient and environmentally responsible energy future (Monaghan et al., 2010; Zhao et al., 2022). Unlike traditional energy systems that often waste a significant portion of energy as heat, polygeneration systems embrace a holistic approach, simultaneously producing multiple energy carriers—electricity, heat, and cooling—from a single energy source. This integrated approach maximizes energy utilization, leading to significantly higher overall efficiencies and reducing reliance on fossil fuels.

The benefits of polygeneration extend far beyond increased efficiency. By decreasing our dependence on fossil fuels, polygeneration systems contribute to lower greenhouse gas emissions and reduced air pollution, playing a crucial role in mitigating the adverse effects of climate change (Farrokhabadi et al., 2017; Nguyen et al., 2021). Moreover, polygeneration systems, often incorporating a diverse portfolio of energy sources, enhance energy security and resilience against disruptions in any single source. This diversified approach provides a more robust and reliable energy infrastructure, reducing vulnerability to price volatility, geopolitical instability, and supply chain disruptions, ultimately leading to a more secure and sustainable energy future.

10.1.4 The Need for Advanced Simulation in Polygeneration

Designing and implementing efficient and cost-effective polygeneration systems are a complex undertaking, requiring a deep understanding of various energy sources, conversion technologies, their dynamic interactions, and varying energy demands (Jawad et al., 2023). This complexity is further amplified by the need to integrate these systems seamlessly into existing energy infrastructures while ensuring grid stability and reliability in the face of fluctuating renewable energy sources. It is within this challenging context that advanced modeling and simulation tools become not merely desirable but indispensable for researchers, engineers, and policymakers alike.

Sophisticated simulation tools provide a virtual proving ground for polygeneration systems, enabling engineers to predict system performance under a wide range of operating conditions, including fluctuations in renewable energy generation, varying energy demands, and potential equipment failures (Miranda et al., 2023; Sattar et al., 2024). By accurately capturing the complex interplay of different system components, energy flows, and external factors, these simulations offer invaluable insights for optimizing system design and operation, leading to more efficient, reliable, and cost-effective solutions. Furthermore, simulation tools empower researchers to explore a multitude of "what-if" scenarios, experimenting with different system configurations, component selections, and control strategies without the risks and costs associated with real-world prototyping. This iterative design process, guided by simulation results, is essential for developing polygeneration systems tailored to specific needs, resource constraints, and operating environments.

10.1.5 Introducing RenewGenix: A Vision for the Future

Designing and implementing efficient and cost-effective polygeneration systems are a complex undertaking. While several commercial software packages exist for energy system simulation (e.g., HOMER, ANSYS, TRNSYS, etc.), they often come with high licensing fees and limited flexibility for customization. This is where RenewGenix comes in.

RenewGenix is envisioned as a pioneering open-source software platform specifically designed to address the critical need for advanced, accessible, and customizable simulation in the field of polygeneration. By embracing an open-source framework, RenewGenix aims to remove the cost barrier and empower a wider range of users—including universities, start-ups, and researchers globally—to contribute to and benefit from its development.

Furthermore, RenewGenix will leverage the power of OpenFOAM, a renowned computational fluid dynamics (CFD) library. This integration is intended to allow for a level of detail and accuracy not often found in more general-purpose energy simulation tools. This high-fidelity modeling could lead to significant cost savings during the design and implementation of polygeneration systems. For instance, by accurately simulating the performance of different system configurations under various operating conditions, engineers would be able to identify and rectify potential design flaws early on, avoiding costly modifications later.

RenewGenix is being envisioned to provide a comprehensive and integrated environment for modeling the complex interactions between diverse energy technologies within polygeneration systems. The goal is for RenewGenix to enable users to:

- **Simulate a Wide Array of Polygeneration Configurations**: Model systems with various combinations of renewable and conventional energy sources, energy storage technologies, and diverse load profiles, tailored to specific applications and research questions.

- **Capture the Dynamic Behavior of These Systems**: Accurately simulate the dynamic interplay between energy carriers, considering fluctuating renewable energy generation, varying load demands, and the unique operational characteristics of different technologies.

- **Evaluate Grid Integration Impacts**: Assess the potential effects of integrating polygeneration systems on the stability and reliability of existing power grids, taking into account factors like voltage fluctuations, frequency variations, and grid protection schemes.

- **Optimize System Design and Operation**: Utilize advanced optimization algorithms to identify the most efficient and cost-effective system configurations and operating strategies, considering user-defined objectives and constraints.

- **Facilitate the Exploration of Emerging Technologies**: Incorporate models for new and innovative renewable energy technologies and energy storage systems, providing a platform for research and development efforts in the field of sustainable energy.

10.2 RenewGenix: Design Goals

10.2.1 Layered Architecture

RenewGenix employs a modular, extensible layered architecture (Fig. 10.1) to ensure flexibility and scalability in modeling complex polygeneration systems. This approach, similar to that in (Svendsen et al., 2023), compartmentalizes functionalities for easier code management, future expansion, and collaborative development. The Data Layer manages all system data, ensuring integrity and consistency. The Modeling Layer defines mathematical models for various components, accurately representing their physical behavior, as exemplified in (Monaghan et al., 2010). The Simulation Engine, powered by OpenFOAM, solves energy flow equations, providing high-fidelity simulations, owing to its sophisticated CFD capabilities, as demonstrated in studies like (Arini et al., 2022; Brown et al., 2020). The Optimization Layer incorporates advanced algorithms, working with user-defined objectives to find optimal solutions, similar to strategies for energy storage system sizing (Jawad et al., 2023). Lastly, the Visualization Layer offers user-friendly interfaces and interactive tools, empowering users to analyze and understand system behavior. This layered architecture makes RenewGenix a powerful, adaptable platform for exploring polygeneration systems.

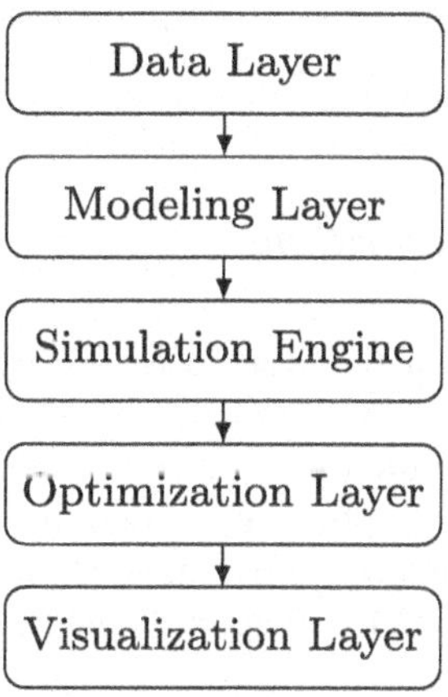

Fig. 10.1 Layered architecture of RenewGenix

10.2.2 Core Features

RenewGenix is designed with a user-centric focus, offering features to streamline polygeneration system simulation and optimization, catering to both experienced OpenFOAM users and newcomers. Its key features include:

- **Flexible Model Representation**: Supports diverse configurations with user-defined energy sources, loads, and components, mirroring the customizability found in platforms like (Svendsen et al., 2023).
- **Advanced Optimization Algorithms**: Integrates powerful algorithms to identify efficient strategies, aligning with optimization approaches used in grid stability enhancement studies (Lai et al., 2024).
- **High-Fidelity Simulation Engine**: Leverages OpenFOAM's CFD capabilities for accurate modeling, as evidenced by studies on Pelton turbine jet detachments (Decaix et al., 2022).
- **Simplified Workflow**: Automates simulation setup and execution, reducing the learning curve and streamlining the process.
- **Seamless OpenFOAM Integration**: Enables users to leverage OpenFOAM's functionalities while benefiting from RenewGenix's user-friendly interface, reflecting trends toward interoperable platforms (Fetsov et al., 2023).
- **Interactive Visualization Tools**: Offers interactive graphs, charts, and 3D visualizations, facilitating clear data representation and promoting effective communication of insights, as emphasized in (Perinov et al., 2021).

10.2.3 Optimization Algorithms

The Optimization Layer, a key part of RenewGenix's layered architecture, is designed to empower users to go beyond mere simulation and delve into the realm of finding the best possible solutions for their polygeneration systems. RenewGenix will integrate a comprehensive suite of optimization algorithms to address a wide range of optimization problems encountered in polygeneration system design and operation. Some of the algorithms planned for integration include:

- **Genetic Algorithms**: Well-suited for exploring large and complex search spaces, genetic algorithms mimic natural selection processes to find optimal solutions. These algorithms are particularly well-suited for multi-objective optimization problems commonly encountered in polygeneration systems, where multiple factors like cost, efficiency, and emissions need to be balanced.
- **Particle Swarm Optimization**: Inspired by the social behavior of bird flocks or fish schools, particle swarm optimization employs a population of candidate solutions that iteratively move toward promising regions of the search space. This approach is effective for finding near-optimal solutions in high-dimensional search

spaces, which are often encountered when optimizing the design and operation of complex polygeneration systems.

- **Simulated Annealing**: Based on the annealing process in metallurgy, simulated annealing uses a probabilistic approach to escape local optima and find globally optimal solutions. This algorithm is particularly useful for optimization problems with complex, nonlinear relationships between variables, as often found in the energy conversion and storage processes within polygeneration systems.

RenewGenix will also provide interfaces for users to define custom objective functions and constraints, allowing for tailored optimization strategies to address specific project requirements. For instance, users might seek to minimize the total cost of a polygeneration system, maximize energy efficiency, or minimize greenhouse gas emissions, subject to constraints such as meeting energy demands, adhering to equipment limitations, or complying with environmental regulations.

10.2.4 Visualization Tools

The Visualization Layer is designed to make the complex data and computations within RenewGenix readily understandable and actionable for users. It provides a suite of interactive visualization tools that empower users to explore data, identify trends, and gain insights into system performance.

- **Time-Series Plots**: These plots will allow users to visualize the dynamic behavior of key system variables over time, such as power generation from renewable sources, energy storage levels, heat production, and energy demand. Users can analyze the interplay between these variables, identify trends, and evaluate the system's response to changes in operating conditions, such as fluctuating renewable energy availability or variations in energy demand.
- **Sankey Diagrams**: Sankey diagrams will provide a clear representation of energy flows within the polygeneration system, illustrating the distribution of energy sources, conversion processes, losses, and end-use demands. This visualization tool enables users to understand the overall energy balance and identify areas for potential efficiency improvements, such as reducing energy losses during conversion processes or optimizing energy distribution to different loads.
- **3D Visualizations**: For components where spatial variations are significant, such as heat exchangers or solar thermal collectors, RenewGenix will offer 3D visualizations. These visualizations will display temperature distributions, flow patterns, and other relevant parameters, providing insights into component performance and enabling users to identify design optimizations or potential issues that might impact the overall performance of the polygeneration system.

The visualization tools will be seamlessly integrated into the RenewGenix user interface, providing a user-friendly and intuitive experience. Users will be able to interact with the visualizations, zooming, panning, and filtering data to gain specific insights into the system's behavior. These interactive capabilities will be crucial for supporting informed decision-making throughout the design, optimization, and operation of polygeneration systems.

10.2.5 Open-Source Framework: Fostering Collaboration and Innovation

RenewGenix wholeheartedly embraces an open-source framework, recognizing its inherent power to accelerate the development and adoption of sustainable energy solutions. This commitment to openness fosters a collaborative and transparent environment that promotes innovation and inclusivity within the renewable energy community. By making the source code freely available, RenewGenix encourages researchers, developers, and practitioners worldwide to contribute their expertise, fostering a vibrant ecosystem of shared knowledge and continuous improvement. This transparency allows for community scrutiny of the underlying models and algorithms, building trust and confidence in the platform's capabilities. Furthermore, the open-source nature of RenewGenix breaks down barriers to entry, making sophisticated simulation tools accessible to a wider audience, including students, independent researchers, and institutions with limited resources. This accessibility democratizes the field of polygeneration system modeling, fostering a more inclusive community and accelerating the pace of innovation. RenewGenix envisions a future where the collective intelligence of a global community drives the development of cutting-edge sustainable energy technologies, and its open-source framework serves as a cornerstone of this vision.

10.2.6 Community Contributions and Quality Control

RenewGenix wholeheartedly embraces an open-source framework, recognizing its inherent power to accelerate the development and adoption of sustainable energy solutions. To ensure the platform's integrity and code quality, we will establish a well-defined process for managing community contributions, guided by the following principles:

- **Version Control System**: RenewGenix's source code will be hosted on a platform like GitHub, utilizing a Git version control system to track changes, manage contributions, and facilitate collaboration. This will allow for transparent tracking of changes, enable easy collaboration between developers, and help resolve merge conflicts.
- **Contribution Guidelines**: Clear and comprehensive guidelines will be provided for submitting code contributions. These guidelines will cover aspects such as coding standards, documentation requirements, and testing procedures, ensuring

that contributions align with the platform's overall design philosophy. This will ensure consistency in coding style and documentation, making it easier for users to understand and utilize the platform's functionalities.

- **Code Review**: All contributions will undergo a thorough code review process by the core development team. This review will assess the code's quality, functionality, and adherence to coding standards. Constructive feedback will be provided to contributors to improve their submissions. This process helps maintain the quality and functionality of the platform while providing valuable feedback to contributors.
- **Testing**: A suite of unit tests and integration tests will be implemented to ensure the functionality and compatibility of contributed modules. Contributors will be encouraged to provide tests for their code, and these tests will be incorporated into the platform's automated testing framework. This will help to ensure that contributions integrate seamlessly with the existing platform and that the overall functionality remains reliable.

By implementing these measures, we aim to foster a collaborative and vibrant community while maintaining the high quality and reliability of the RenewGenix platform.

10.2.7 High-Performance Computing for Scalability

To effectively tackle the computational demands of simulating complex polygeneration systems, particularly those involving high-fidelity CFD models, RenewGenix leverages the power of high-performance computing (HPC). This strategic approach ensures that the platform can scale to handle intricate models with a multitude of components and variables, allowing for the analysis of large-scale and realistic polygeneration scenarios. RenewGenix employs parallel processing techniques, specifically Message Passing Interface (MPI) and shared-memory threading with OpenMP, to distribute computations across multiple processors and computer nodes. This parallelization strategy, visually represented in Fig. 10.2, significantly accelerates simulation runtimes, enabling users to explore a wider range of design options and operating conditions in a timely manner. Furthermore, RenewGenix integrates GPU acceleration to further enhance computational speed and efficiency. By harnessing the parallel processing power of graphics processing units (GPUs), particularly beneficial for computationally intensive CFD simulations, RenewGenix can tackle complex fluid flow and heat transfer phenomena with remarkable speed and accuracy. This integration of HPC techniques, mirroring the methodologies employed in studies on simulating wave interactions with floating offshore structures (Ding et al., 2024) and performing high-fidelity simulations of wind turbines (Zhangaskanov et al., 2022), makes RenewGenix a truly scalable platform, capable of meeting the computational demands of cutting-edge research and real-world engineering projects.

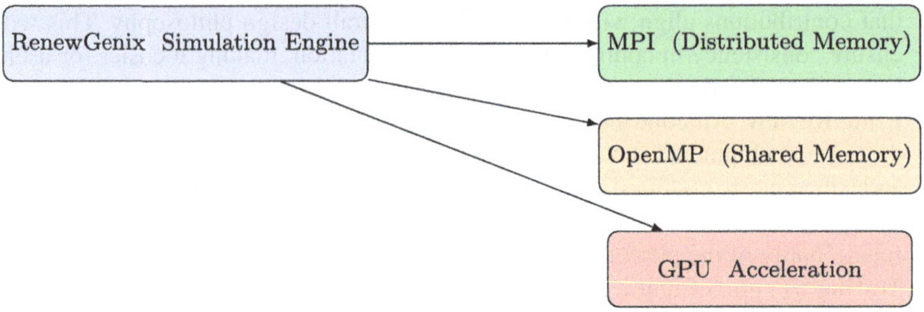

Fig. 10.2 RenewGenix leverages parallel computing techniques for high-performance simulations. MPI enables distribution across multiple nodes, OpenMP utilizes multi-core processors, and GPUs provide specialized hardware acceleration

10.3 RenewGenix Framework: Detailed File Structure

The RenewGenix project adheres to a meticulously structured and modular file organization to promote code readability, maintainability, and extensibility. This systematic approach, visually represented in the figures throughout this section, streamlines navigation for developers, facilitates seamless integration of new modules, and elucidates the intricate relationships between the platform's various components.

10.3.1 Top-Level Structure

At the top level, the RenewGenix project directory is meticulously divided into essential subdirectories, each serving a distinct purpose to ensure a well-defined separation of concerns (Figs. 10.3 and 10.4).

- **docs/:** This directory serves as the central repository for all project documentation, including comprehensive user guides that provide step-by-step instructions for using the platform, detailed API documentation for developers integrating RenewGenix into their workflows, and developer notes to facilitate collaboration and code maintenance.
- **'src/':** This directory forms the core of the RenewGenix platform, containing the complete source code.
- **'tests/':** To uphold high code quality and ensure the reliability of RenewGenix, a dedicated 'tests/' directory houses a comprehensive suite of unit tests and integration tests. Unit tests focus on verifying the functionality of individual code units, while integration tests assess the interaction and compatibility between different modules.

The remaining top-level directories contribute to the efficient development and deployment of RenewGenix:

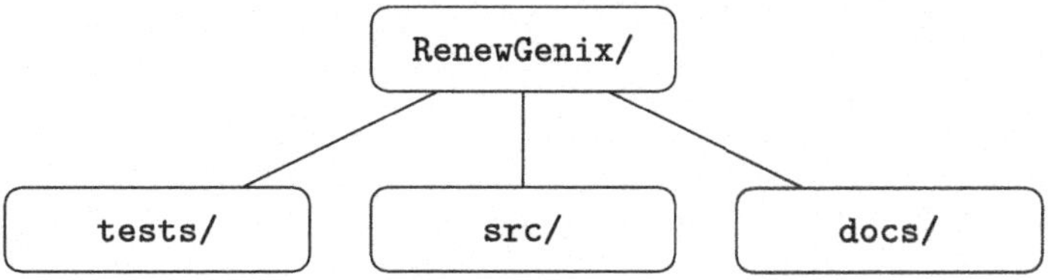

Fig. 10.3 Top-level structure of the RenewGenix project directory (Part A)

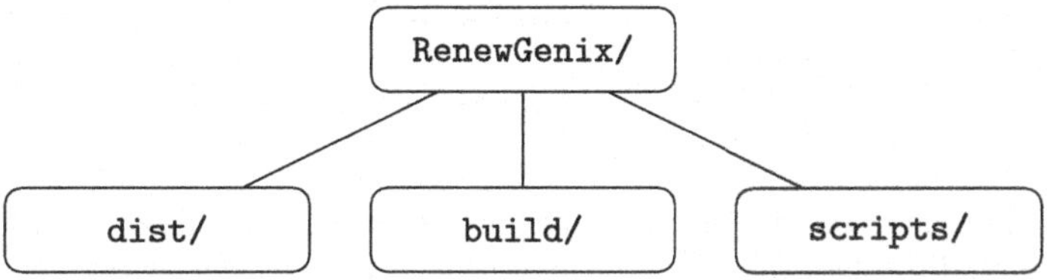

Fig. 10.4 Top-level structure of the RenewGenix project directory (Part B)

- **scripts/:** This directory houses utility scripts essential for building the software from source code, running automated tests, and deploying the software to various environments, streamlining the development workflow and ensuring consistency across different stages.
- **'build/':** The 'build/' directory serves as a temporary storage space during the compilation process, holding intermediate build files and artifacts generated by the build system. This separation keeps the source code directory clean and organized.
- **'dist/':** This directory contains the final, distributable packages of the RenewGenix software, prepared for installation on different operating systems and platforms. These pre-packaged bundles facilitate easy distribution and deployment to end-users.

10.3.2 Structure of the src/ Directory

The src/ directory constitutes the core of RenewGenix, housing the source code for all its modules and functionalities. Its structure reflects the layered architecture, providing a clear organizational framework for the platform's diverse capabilities.
Core Functionalities (core/) The core/ directory (Fig. 10.5) contains the fundamental building blocks of RenewGenix, implementing the essential layers of its architecture:

- **simulation/:** This subdirectory houses the implementation of the simulation engine, including the crucial interface for interacting with the OpenFOAM CFD library.

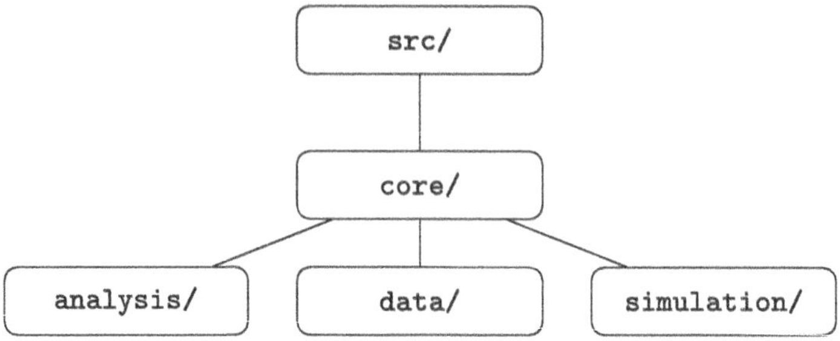

Fig. 10.5 Structure of the src/core directory

- **data/:** The data/ subdirectory manages all aspects of data within RenewGenix, including data input/output operations, defining data structures for storing and manipulating simulation data, and ensuring data integrity and consistency throughout the platform.
- **analysis/:** This subdirectory provides a toolkit for analyzing simulation results, including functions and classes for calculating key performance metrics, generating insightful reports, and equipping users with tools to effectively interpret simulation outcomes.

Modular Components (modules/) The modules/ directory houses individual modules representing components found in polygeneration systems. This modularity facilitates extensibility, enabling developers to add new technologies and component models easily as the platform evolves.
Photovoltaic Systems (pv/) This module (Figs. 10.6 and 10.7) encapsulates functionalities related to photovoltaic (PV) systems, containing models for different types of PV panels, inverters, Maximum Power Point Tracking (MPPT) algorithms, and mounting systems. It is further organized into subdirectories that reflect the different aspects of PV system modeling and analysis:

- **panel_models/:** Contains models for various PV panel technologies, including monocrystalline, polycrystalline, thin-film, and emerging technologies. This subdirectory may include models that simulate the electrical characteristics of PV panels under varying irradiance and temperature conditions, accounting for factors like efficiency, degradation, and shading effects.
- **inverter_models/:** Includes models for different types of inverters used in PV systems, such as string inverters, microinverters, and central inverters. These models simulate the conversion of DC power from PV panels to AC power for grid connection or local consumption, considering efficiency, power output limitations, and control functionalities.
- **mppt_algorithms/:** Houses implementations of various Maximum Power Point Tracking (MPPT) algorithms, which are crucial for optimizing the power output of PV systems. This subdirectory may include algorithms like Perturb and Observe

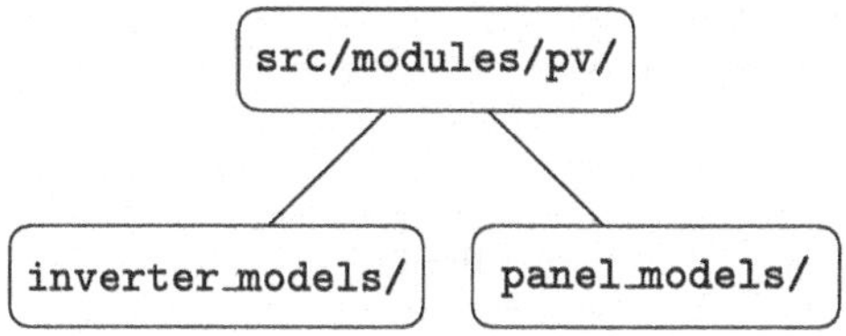

Fig. 10.6 Structure of the src/modules/pv directory (Part A)

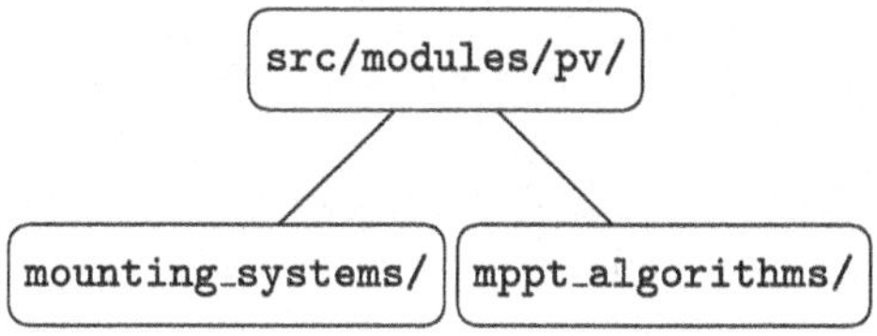

Fig. 10.7 Structure of the src/modules/pv directory (Part B)

(P&O), Incremental Conductance (INC), and more advanced techniques like fuzzy logic or artificial neural network-based MPPT.

- **mounting_systems/:** Contains models for different types of PV mounting systems, such as fixed-tilt, single-axis tracking, and dual-axis tracking. These models simulate the influence of mounting angles and tracking mechanisms on the solar irradiance received by PV panels, considering factors like shading, row spacing, and ground reflectance.

Wind Energy Systems (wind/)

This module (Figs. 10.8, 10.9, and 10.10) includes models for wind energy systems, divided into submodules for horizontal-axis wind turbines (HAWT/) and vertical-axis wind turbines (VAWT/):

- **HAWT/:** This submodule focuses on horizontal-axis wind turbines (HAWTs), the most common type of wind turbine. It is further organized into subdirectories reflecting different aspects of HAWT modeling and analysis:

• **aerodynamics/:** Contains aerodynamic models for HAWT blades, including blade element momentum theory (BEMT) implementations and computational fluid dynamics (CFD) models for high-fidelity simulations.
• **control/:** Includes control algorithms for essential aspects of HAWT operation, such as pitch control (adjusting blade angle for optimal power output and speed regulation) and yaw control (rotating the nacelle to face the wind).
• **structural/:** Contains structural models and analysis tools for HAWT components (tower, blades, support structures), which might include finite element analysis (FEA) models for assessing structural integrity under various loading conditions.

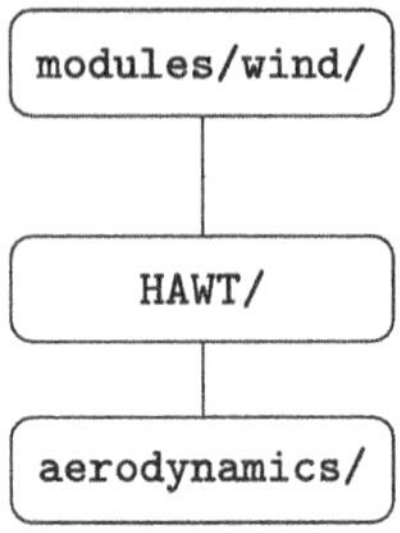

Fig. 10.8 Detailed structure of the modules/wind/HAWT subdirectory

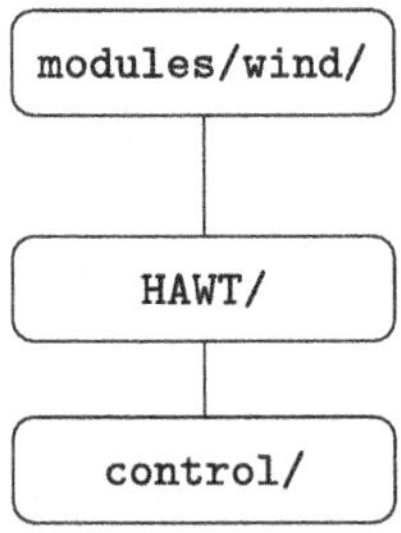

Fig. 10.9 Detailed structure of the modules/wind/HAWT subdirectory

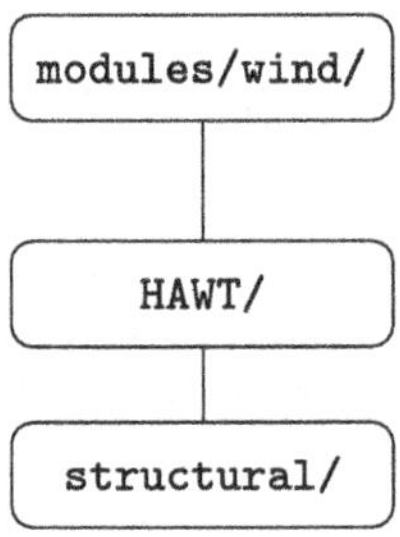

Fig. 10.10 Detailed structure of the modules/wind/HAWT subdirectory

– **VAWT/:** This submodule focuses on vertical-axis wind turbines (VAWTs), mirroring the HAWT/ submodule's organization:

- **aerodynamics/:** Contains aerodynamic models for VAWT blades, such as double multiple streamtube models for simplified analysis and CFD models for more detailed simulations.
- **design/:** Includes tools and calculations specifically for designing VAWT rotors, blades, and support structures, potentially incorporating optimization algorithms for determining optimal blade shapes and configurations.

Thermal Energy Systems ('thermal/') The 'thermal/' module (Figs. 10.11, 10.12, 10.13, and 10.14) encompasses models for thermal energy systems, including components like solar thermal collectors, boilers, and heat exchangers. It is further organized into subdirectories:

- **'solar_thermal_collectors/':** Contains models for various types of solar thermal collectors:

  - **'flat_plate/':** Models for flat-plate collectors, the most common type of solar thermal collector, suitable for applications like domestic hot water heating and space heating.
  - **'evacuated_tube/':** Models for evacuated tube collectors, which offer higher efficiency and better performance in cold climates compared to flat-plate collectors, often used for higher temperature applications.
  - **'cpc/':** Models for compound parabolic concentrator (CPC) collectors, which utilize non-imaging optics to concentrate sunlight onto a receiver tube, achieving higher temperatures and efficiency than flat-plate collectors.
  - **'parabolic_trough/':** Models for parabolic trough collectors, which use curved mirrors to focus sunlight onto a receiver tube running along the focal line of the parabola, suitable for industrial process heat and electricity generation.
  - **'dish_stirling/':** Models for dish Stirling systems, which combine a parabolic dish concentrator with a Stirling engine to convert concentrated solar energy into electricity, achieving high concentration ratios and efficiency.

- **'boilers/':** Includes models for different types of boilers used in thermal energy systems:

  - **'gas_fired/':** Models for gas-fired boilers, which use natural gas or propane as fuel, commonly used for space heating and hot water production.
  - **'biomass/':** Models for biomass boilers, which burn wood pellets, chips, or other biomass materials, providing a renewable source of heat.
  - **'electric/':** Models for electric boilers, which use electricity to heat water, often used in combination with renewable energy sources or for specific applications where direct combustion is not feasible.

- **'heat_exchangers/':** Contains models for various types of heat exchangers used for transferring thermal energy between different fluids or mediums:

  - **'shell_and_tube/':** Models for shell-and-tube heat exchangers, the most common type, where one fluid flows through a bundle of tubes enclosed within a shell, and the other fluid flows through the shell.
  - **'plate/':** Models for plate heat exchangers, which use a series of thin, corrugated plates to create channels for fluid flow, providing a compact and efficient design.

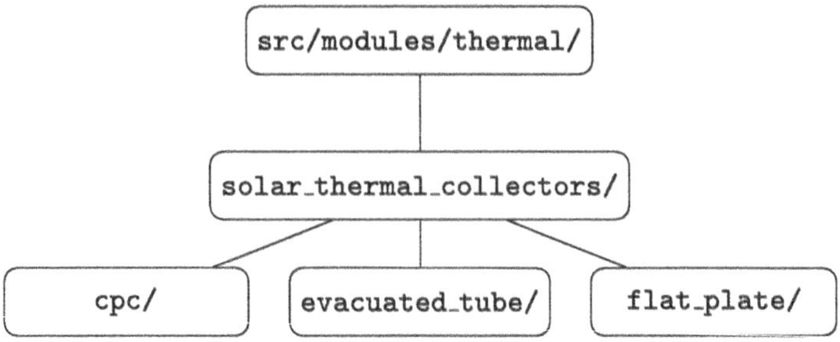

Fig. 10.11 Structure of the src/modules/thermal/solar_thermal_collectors directory (Part A)

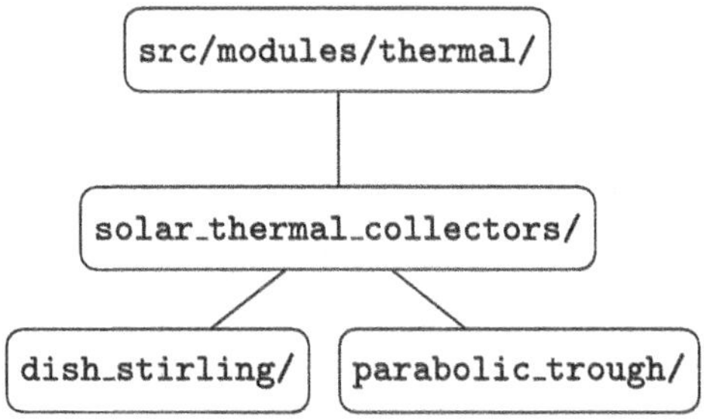

Fig. 10.12 Structure of the src/modules/thermal/solar_thermal_collectors directory (Part B)

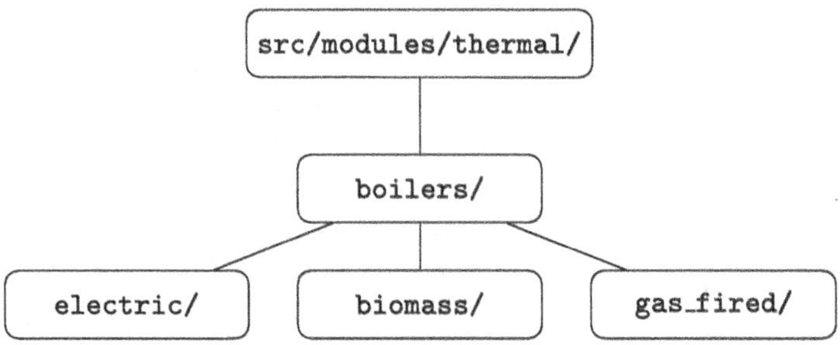

Fig. 10.13 Structure of the src/modules/thermal/boilers directory

- **'air-cooled/':** Models for air-cooled heat exchangers, which use air as the cooling medium, typically employed in applications where water availability is limited or for heat dissipation purposes.

Energy Storage ('storage/') The 'modules/storage/' directory (Figs. 10.15, 10.16 and 10.17) focuses on models and algorithms for different energy storage technologies, a crucial aspect of polygeneration systems, especially those with a high penetration of intermittent renewable sources:

- **'batteries/':** This submodule focuses on battery energy storage systems, with subdirectories reflecting various aspects of battery modeling:

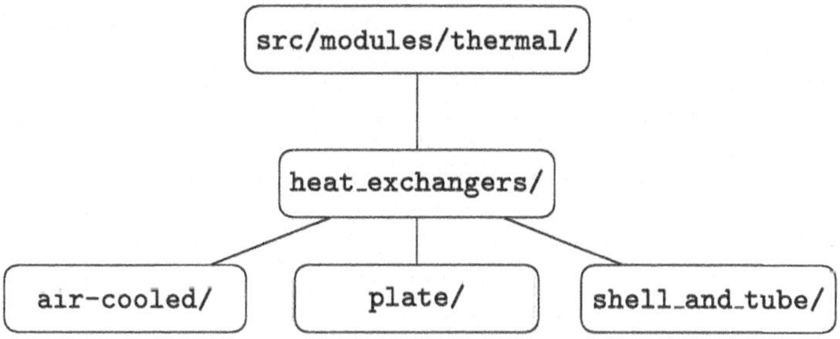

Fig. 10.14 Structure of the src/modules/thermal/heat_exchangers directory

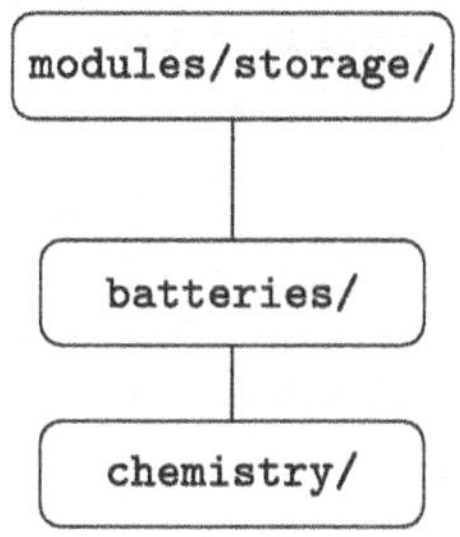

Fig. 10.15 Detailed structure of the modules/storage/batteries subdirectory

- **'chemistry/':** Contains models for different battery chemistries, including lithium-ion, lead-acid, flow batteries, and emerging technologies. This subdirectory may include models for charging/discharging characteristics, degradation mechanisms, and thermal behavior.
- **'management_system/':** Includes models for battery management systems (BMSs) responsible for monitoring and controlling battery operation to ensure safe and optimal performance. This subdirectory may contain algorithms for battery state estimation, charge control, cell balancing, and fault detection.

– **'pumped_hydro/':** This submodule contains models for pumped hydro energy storage (PHES) systems, which store energy by pumping water uphill to a reservoir and release it by generating electricity through turbines when needed. This subdirectory may include models for pump/turbine characteristics, reservoir dynamics, and overall system efficiency.

Battery models in RenewGenix will incorporate critical constraints, including charge and discharge power limits and initial and final energy constraints, as highlighted in studies like (Nezhad et al., 2022). These constraints ensure realistic representation of battery operation within polygeneration systems.

Power Electronics ('power_electronics/') This module houses models for power electronics components prevalent in renewable energy systems, with submodules for:

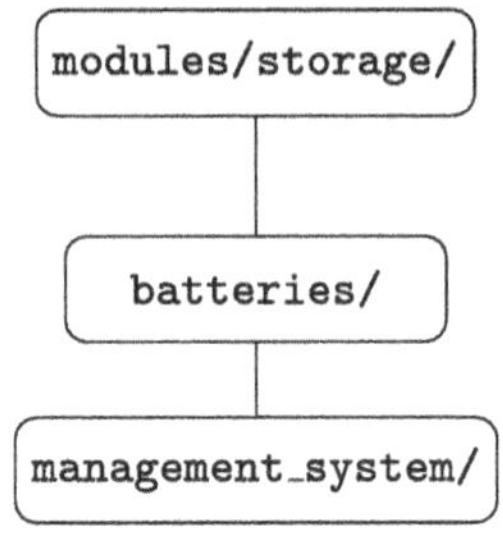

Fig. 10.16 Detailed structure of the modules/storage/batteries subdirectory

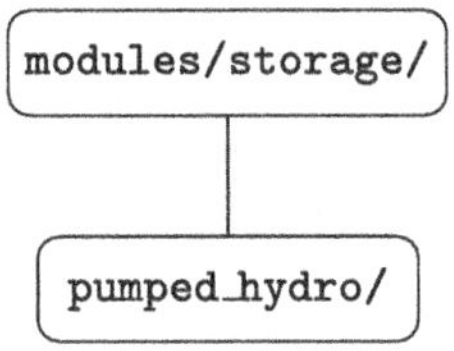

Fig. 10.17 Detailed structure of the modules/storage subdirectory

- **'inverters/':** Contains models for inverters, which are essential for converting DC power from sources like PV panels and batteries to AC power for grid connection or local use. This subdirectory may include models for different inverter topologies, control strategies, and efficiency characteristics.
- **'solar_regulators/':** Includes models for solar charge controllers, also known as solar regulators, which manage the charging of batteries from PV panels, ensuring safe and efficient operation. This subdirectory may include models for different charge controller types, such as PWM and MPPT, and their specific algorithms.
- **'other_devices/':** This subdirectory can accommodate models for other power electronics components used in polygeneration systems, such as rectifiers, converters, and power factor correction (PFC) devices, allowing for a comprehensive representation of the electrical power flow and conversion within the system.

Utilities, External Interfaces, and User Interface The remaining directories within 'src/' provide essential functionalities for integration and user interaction (see Fig. 10.18):

- **'utils/':** This directory contains utility functions and classes reusable across different modules, promoting code efficiency and reducing redundancy. It may include functions for mathematical operations, data processing, file handling, and other general-purpose functionalities.
- **'external/':** This directory manages interfaces and wrappers for external libraries and tools, bridging RenewGenix with the wider ecosystem of software resources:

- **'OpenFOAM/':** This subdirectory contains the interface code that enables RenewGenix to leverage the OpenFOAM CFD library, handling data exchange between the platforms for high-fidelity simulations while maintaining a user-friendly environment within RenewGenix.

- **'meteo/':** This subdirectory provides interfaces for accessing and processing meteorological data from external sources, including weather data APIs, historical databases, and other relevant sources. This integration is crucial for accurate simulations, especially for weather-dependent renewable energy systems.

The 'external/OpenFOAM/' subdirectory plays a crucial role in enabling RenewGenix to leverage the advanced CFD capabilities of OpenFOAM. This subdirectory contains a dedicated interface module responsible for managing the seamless data exchange between the two platforms. This interface module handles several key tasks:

- **Data Preparation**: The interface module takes the system configuration defined in RenewGenix (component types, properties, connections, operating parameters) and translates this information into OpenFOAM-compatible input files. This includes generating the computational mesh, defining boundary conditions, and specifying material properties for the CFD simulation.
- **Simulation Execution**: The interface module initiates and controls the OpenFOAM simulation, handling the execution of the appropriate OpenFOAM solvers based on the system configuration.
- **Result Retrieval**: Upon completion of the OpenFOAM simulation, the interface module retrieves the relevant results, such as velocity fields, pressure distributions, temperature profiles, and other flow-related data.
- **Data Conversion**: The interface module converts the OpenFOAM results into a format readily usable by RenewGenix. This may involve data extraction, interpolation, or post-processing to align with the data structures and analysis tools within RenewGenix.

This bidirectional data flow between RenewGenix and OpenFOAM allows users to seamlessly incorporate high-fidelity CFD simulations into their polygeneration system models. The interface module acts as a bridge, shielding users from the complexities of OpenFOAM while enabling them to benefit from its advanced simulation capabilities. This integration significantly enhances the accuracy and detail of RenewGenix's simulations, leading to more reliable performance predictions and optimization results.

- **'ui/':** This directory houses the code for the user interface, which could be a graphical user interface (GUI) or a command-line interface (CLI). The user interface allows users to define simulation parameters, input system configurations, execute simulations, visualize results, and perform other essential tasks.

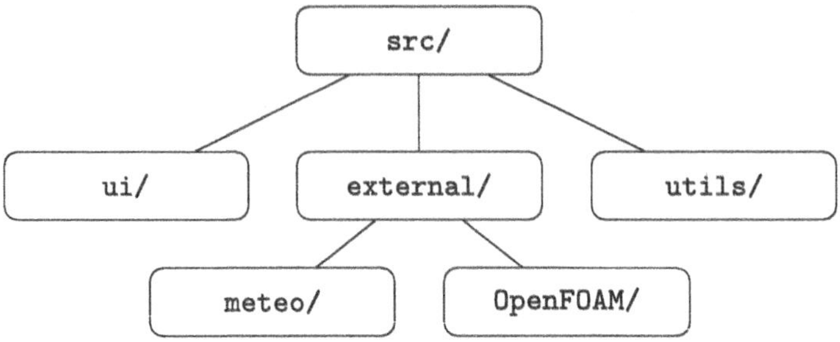

Fig. 10.18 Structure of the 'src' directory: utilities, external interfaces, and user interface

10.4 RenewGenix: Potential Applications and Impact

RenewGenix, with its comprehensive modeling capabilities, user-friendly interface, and open-source nature, is poised to have a significant impact on the design, optimization, and analysis of polygeneration systems, accelerating their adoption and contributing to a more sustainable energy future. This section explores the potential applications of RenewGenix across various energy system paradigms, highlighting its capabilities in addressing key challenges and driving innovation in the field of sustainable energy.

10.4.1 Hybrid Renewable Energy Systems and Energy Storage

The increasing demand for reliable and sustainable energy has led to a surge in the development of hybrid renewable energy systems that combine multiple renewable sources, such as solar PV, wind turbines, biomass generators, and hydroelectric power. These hybrid systems leverage the complementary strengths of different technologies, mitigating the inherent intermittency of individual renewable sources and ensuring a more consistent and reliable energy supply. RenewGenix provides a robust platform for designing and optimizing such hybrid systems. Its flexible model representation allows users to easily integrate models of different renewable energy technologies, along with energy storage systems, load profiles, and other relevant components. RenewGenix can then be used to simulate the performance of these hybrid systems under varying operating conditions, considering factors like weather patterns, energy demand fluctuations, and component performance characteristics. The platform's advanced optimization algorithms can then be employed to identify optimal system configurations, considering user-defined objectives such as maximizing energy output, minimizing cost, or minimizing environmental impact. For example, RenewGenix could be used to design a hybrid system for a remote community that combines solar PV, wind turbines, and battery storage. The platform

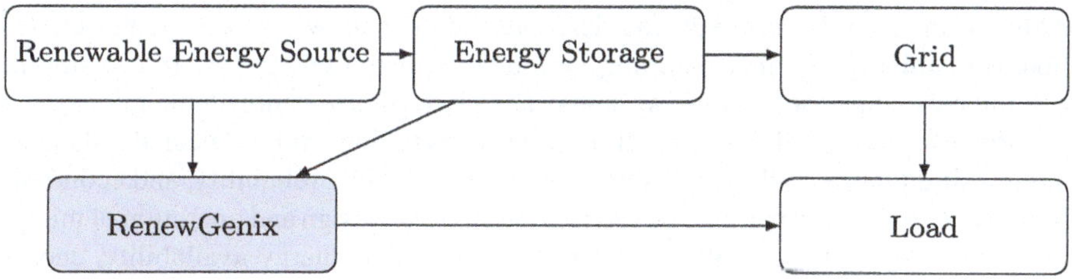

Fig. 10.19 Energy storage integration and management within a renewable energy system using RenewGenix

can simulate the system's performance throughout the year, considering variations in solar irradiance, wind speeds, and energy demand. The optimization algorithms can then be used to determine the optimal sizing of the PV array, wind farm, and battery storage system, ensuring a reliable and cost-effective energy supply for the community. Energy storage is a crucial element in modern energy systems, particularly those with high penetration of intermittent renewable sources. Batteries, thermal energy storage, pumped hydro storage, and other emerging technologies play a vital role in smoothing out power fluctuations, improving grid stability, and enabling the integration of larger amounts of renewable energy (Fig. 10.19). RenewGenix offers a comprehensive suite of tools for simulating and optimizing energy storage systems within polygeneration systems. Users can incorporate detailed models of various storage technologies, considering factors like storage capacity, charging/discharging rates, efficiency, and degradation. The platform can simulate the dynamic behavior of these storage systems in response to fluctuations in renewable energy generation and energy demand, enabling users to assess their impact on overall system performance and grid stability. Furthermore, RenewGenix's optimization algorithms can be used to determine optimal control strategies for energy storage systems, maximizing their effectiveness in meeting specific objectives. For instance, the platform could be used to optimize the operation of a battery energy storage system connected to a wind farm, aiming to minimize energy curtailment during periods of high wind generation and maximize the use of stored energy during periods of low wind generation, thereby ensuring a more stable and predictable power output to the grid.

10.4.2 Microgrid Design and Optimization

Microgrids represent localized energy systems capable of operating independently or in conjunction with the main grid, gaining traction as a crucial step toward a more decentralized and resilient energy infrastructure. These self-sufficient systems often integrate a mix of renewable energy sources, energy storage, and controllable loads, promising enhanced reliability, reduced energy losses, and increased energy independence. RenewGenix provides a versatile platform for designing, simulating, and analyzing microgrids. Its flexible model representation allows users to create detailed models of microgrids, incorporating various distributed energy resources, storage technologies, loads, and control systems. The platform can simulate the microgrid's performance under different operating modes, including grid-connected, islanded, and transition modes, allowing users to assess its stability, reliability, and economic viability. RenewGenix can also be used to optimize the design and operation of microgrids, considering factors like load profiles, renewable energy availability, energy storage capacity, and economic constraints. The platform's optimization algorithms can assist in determining the optimal mix of energy sources, storage capacity, and control strategies to meet the specific needs of a microgrid, maximizing its efficiency, reliability, and cost-effectiveness. For instance, RenewGenix could be used to design and optimize a microgrid for a university campus that incorporates solar PV, a combined heat and power (CHP) plant, battery storage, and smart load management. The platform can simulate the microgrid's operation, considering daily and seasonal variations in energy demand, solar irradiance, and electricity prices. The optimization algorithms can then be used to determine the best operating strategies for the microgrid, aiming to minimize energy costs, reduce greenhouse gas emissions, and ensure reliable power supply to the campus.

10.4.3 Industrial Polygeneration Systems

Polygeneration, the simultaneous generation of multiple energy forms (e.g., electricity, heat, cooling) from a single energy source, offers significant potential for enhancing energy efficiency, reducing costs, and minimizing environmental impact in industrial settings. Industrial facilities often have diverse energy demands, and polygeneration systems can effectively utilize waste heat and other byproducts of industrial processes to generate additional valuable energy forms (Fig. 10.20). RenewGenix provides a powerful tool for designing and optimizing polygeneration systems tailored to the specific needs of industrial applications. Its flexible framework allows users to model various industrial processes and their energy demands, along with different polygeneration technologies and configurations. The platform can simulate the performance of these systems under varying operating conditions, assessing their impact on energy efficiency, emissions, and overall cost. The optimization algorithms within RenewGenix can then be employed to identify optimal designs and operating strategies for polygeneration systems, aiming to maximize energy recovery, reduce reliance on external energy sources, and minimize environmental footprint. For example, RenewGenix could be used to design and optimize a polygeneration system for a food processing plant that utilizes waste heat from production processes to generate electricity and provide process heating and cooling, leading to significant energy savings and reduced greenhouse gas emissions.

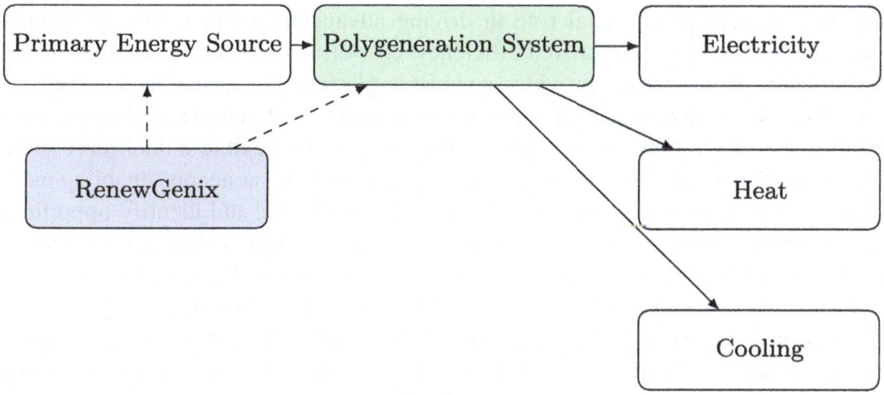

Fig. 10.20 Polygeneration system producing multiple energy outputs from a single energy source, modeled and optimized using RenewGenix

10.4.4 Real-Time Capabilities

To enable more dynamic and responsive polygeneration system operation, RenewGenix will incorporate features to support real-time data acquisition and control. These features will be designed to facilitate the integration of RenewGenix with real-world systems and enable intelligent energy management decisions. Some of the key features planned for this area include:

- **Real-Time Data Acquisition**: Interfaces to connect with sensors and data sources providing real-time information about energy generation, consumption, storage levels, and grid conditions.
- **Dynamic Control Algorithms**: Support for advanced control algorithms, such as rule-based control, model predictive control, and reinforcement learning, to enable dynamic adjustments to system operation in response to real-time changes.
- **Visualization and Monitoring**: Real-time dashboards and visualizations to provide operators with up-to-date information and insights into system performance.

These advancements would transform RenewGenix into a powerful tool for real-time energy management in polygeneration systems, enabling optimal control, grid stability, and enhanced system efficiency.

10.4.5 Research and Development of New Renewable Energy Technologies

The field of renewable energy is constantly evolving, with new technologies and innovative concepts emerging to harness energy from various natural sources. Research and development play a vital role in driving advancements in renewable energy technologies, leading to improved efficiency, cost-effectiveness, and environmental performance. RenewGenix provides a valuable platform for researchers and developers to explore the potential of new renewable energy technologies and assess their viability for integration into polygeneration systems. Its flexible model representation allows for the incorporation of novel technologies and concepts, enabling users to simulate their performance, understand their limitations, and identify opportunities for optimization. For instance, researchers could use RenewGenix to investigate the performance of novel solar thermal collectors, advanced battery chemistries, or innovative wave energy converters, simulating their integration into polygeneration systems and evaluating their potential to enhance energy efficiency and reduce environmental impact. By providing a robust and adaptable framework for modeling and simulation, RenewGenix facilitates the development of next-generation renewable energy technologies, accelerating the transition toward a more sustainable energy future.

10.5 Comparison with Existing Open-Source Tools

RenewGenix aims to offer a distinct set of capabilities compared to other open-source energy system simulation tools like HOMER and TRNSYS. While HOMER is primarily focused on microgrid design and optimization, and TRNSYS excels in detailed building energy simulations, RenewGenix occupies a unique niche with its specific emphasis on polygeneration systems. It will also leverage the power of OpenFOAM for high-fidelity CFD simulations, providing a level of detail and accuracy not often found in more general-purpose energy simulation tools. RenewGenix's open-source framework also fosters a collaborative environment for community development and customizability, empowering users to contribute to the platform's growth and tailor it to their specific needs.

However, as RenewGenix is currently under development, comprehensive comparisons with other platforms are not yet possible. Future publications will include a detailed analysis of RenewGenix's performance and how it compares to existing tools like HOMER and TRNSYS. This analysis will provide valuable insights for potential users and developers.

10.6 Preliminary Performance Evaluation

While a comprehensive performance evaluation is currently underway, preliminary tests using a simplified polygeneration system model demonstrate the potential benefits of RenewGenix's parallel processing capabilities. This simplified model, consisting of a solar PV array, battery storage, and a heat exchanger, was simulated over a 24-hour period. Utilizing OpenMP for shared-memory parallelization on a multi-core processor resulted in a 35% reduction in simulation time compared to a single-threaded execution. These early results highlight the potential of RenewGenix to handle computationally intensive simulations efficiently, enabling users to explore a wider range of design options and operating conditions in a time-effective manner. More comprehensive benchmark results, showcasing the platform's scalability and performance with increasing model complexity, will be presented in future publications.

10.7 Development Roadmap and Next Steps

The development of RenewGenix will follow a phased approach, characterized by a commitment to community engagement and continuous improvement. Each phase will build upon the previous one, expanding the platform's capabilities, refining its functionalities, and ensuring its relevance to the evolving landscape of sustainable energy technologies.

10.7.1 Phased Development Approach

The development roadmap for RenewGenix outlines three key phases:

- **Phase 1: Core Framework Development**: This initial phase will focus on establishing the foundation of the platform, including the implementation of the layered architecture, core modules for data handling, simulation, basic optimization, and visualization, and an initial set of component models for PV, wind, and basic thermal systems. A functional prototype demonstrating these core capabilities will be developed and released as open-source code, allowing early adopters and the community to provide feedback and contribute to the platform's evolution.
- **Phase 2: Model Library Expansion and Refinement**: In this phase, the focus will shift to expanding the component model library to encompass a wider range of technologies, including advanced PV, wind, detailed thermal, and storage models. Existing models will be refined based on feedback from the community and the latest research findings. The optimization algorithms and visualization tools will also be enhanced, providing users with more sophisticated capabilities for analyzing and optimizing polygeneration systems.
- **Phase 3: User Interface and HPC Integration**: The final phase will concentrate on developing a user-friendly graphical interface (GUI) to make RenewGenix more accessible to a broader range of users. Integration with high-performance computing (HPC) resources will also be a priority, allowing for the efficient simulation of large-scale and complex polygeneration systems. This phase will culminate in the release of a fully-featured and user-friendly version of RenewGenix, ready for widespread adoption by researchers, engineers, and policymakers.

10.7.2 Validation Strategy

A robust validation strategy is essential to ensure the accuracy and reliability of RenewGenix. The validation process will encompass three key areas:

- **Component Model Validation**: Individual component models will be validated against experimental data, established theoretical models, or results from other trusted simulation tools. For instance, PV panel models will be compared against manufacturer specifications and performance data, while heat exchanger models will be validated against experimental measurements of heat transfer rates.
- **System-Level Validation**: Complete polygeneration system models will be tested against real-world data from existing systems or detailed case studies. Key performance indicators, such as energy output, efficiency, emissions, and operating costs, will be compared to assess the accuracy and predictive capability of the simulations.
- **Optimization Algorithm Validation**: The performance and accuracy of the implemented optimization algorithms will be evaluated using benchmark optimization problems and by comparing results with alternative optimization techniques. This will ensure that the algorithms consistently find optimal or near-optimal solutions.

10.8 Conclusion

RenewGenix is not just a software platform; it's a vision for a future powered by collaborative innovation in the field of polygeneration. By embracing an open-source framework and planning to utilize the power of OpenFOAM, RenewGenix offers a unique opportunity to democratize access to advanced simulation tools, potentially accelerating the development and adoption of sustainable energy solutions globally.

This chapter has laid out the blueprint for RenewGenix, detailing its architecture, intended core features, and wide-ranging applications. But the true potential of RenewGenix lies in its open and collaborative nature. We envision a vibrant community of researchers, engineers, and industry professionals actively contributing to the platform's growth. Imagine a future where:

- Students could use RenewGenix to explore and learn about the intricacies of polygeneration systems, contributing to the next generation of clean energy experts.
- Researchers could seamlessly integrate their novel energy models and algorithms into RenewGenix, accelerating scientific discovery and the development of cutting-edge technologies.
- Engineers could leverage the platform's high-fidelity simulations to design and optimize real-world polygeneration systems, leading to more efficient and cost-effective energy solutions for communities and industries worldwide.

The preliminary performance results and the ongoing validation efforts underscore the platform's potential to become a powerful and reliable tool for advancing sustainable energy solutions. By fostering a collaborative open-source community, we aim to accelerate the development, adoption, and impact of RenewGenix, driving progress toward a cleaner, more efficient, and sustainable energy future.

References

Alrai, M., Abuwardeh, M., Al-Ghzaiwat, M., & As'ad, S. (2024). Assessment of the penetration impact of renewable-rich electrical grids: The Jordanian grid as a case study. *MRS Energy Sustainability, 11*(1).

Arini, N. R., Turnock, S. R., & Tan, M. (2022). The effect of trailing edge profile modifications to fluid-structure interaction of a vertical axis tidal turbine blade. *International Journal of Renewable Energy Development, 11*(3), 725–735.

Brown, S. A., Ransley, E. J., Zheng, S., Xie, N., Howey, B., & Greaves, D. M. (2020). Development of a fully nonlinear, coupled numerical model for assessment of floating tidal stream concepts. *Ocean Engineering*, 218.

Decaix, J., & Münch-Alligné, C. (2022). Geometry, mesh and numerical scheme influencing the simulation of a Pelton jet with the OpenFOAM toolbox. *Energies, 15*(19), 7451.

Ding, H., Qiang, C., & Zang. (2024, June). Numerical simulation of wave interactions with floating offshore renewable energy structures: A comparative study between a particle-based PIC model and OpenFOAM. *Journal of Fluids and Structures, 126*, 104092.

Esmaeel Nezhad, A., Jowkar, S., Tavakkoli Sabour, T., Rahimi, E., Ghanavati, F., & Esmaeilnezhad, F. (2024). A short-term wind-hydrothermal operational framework in the presence of pumped-hydro storage. *e-Prime—Advances in Electrical Engineering, Electronics and Energy, 8*, 100577.

Farrokhabadi, M., Cañizares, C. A., & Bhattacharya, K. (2017). Frequency control in isolated/islanded microgrids through voltage regulation. *IEEE Transactions on Smart Grid, 8*(3), 1185–1194.

Fetsov, S. S., & Lutsenko, N. A. (2023). A novel computational model and OpenFOAM solver for simulating thermal energy storages based on granular phase change materials: Advantages and applicability. *Journal of Energy Storage, 65*, 107294.

Groppi, D., Garcia, D. A., Basso, G. L., Cumo, F., & De Santoli, L. (2018). Analysing economic and environmental sustainability related to the use of battery and hydrogen energy storages for increasing the energy independence of small islands. *Energy Conversion and Management, 177*, 64–76.

Herc, L., Pfeifer, A., Duić, N., & Wang, F. (2022). Economic viability of flexibility options for smart energy systems with high penetration of renewable energy. *Energy, 252*, 123739.

Javadi, M. S., Anvari-Moghaddam, A., Guerrero, J. M., Nezhad, A. E., Lotfi, M., & Catalão, J. P. S. (2019). Optimal operation of an energy hub in the presence of uncertainties. In *2019 IEEE International Conference on Environment and Electrical Engineering and 2019 IEEE Industrial and Commercial Power Systems Europe (EEEIC/I&CPS Europe)* (pp. 1–4). IEEE.

Javadi, M. S., Nezhad, A. E., Gough, M., Lotfi, M., Anvari-Moghaddam, A., Nardelli, P. H., Sahoo, S., & Catalão, J. P. (2021). Conditional value-at-risk model for smart home energy management systems. *e-Prime—Advances in Electrical Engineering, Electronics and Energy, 1*, 100006.

Jawad, A., Ahmed, K. I., Mahmud, & I., Nahid-Al-Masood. (2023, November). Optimal supercapacitor energy storage system sizing for frequency control in a large-scale photovoltaic integrated grid. In *2023 International Conference on Technology and Policy in Energy and Electric Power (ICT-PEP)* (pp. 52–57). IEEE.

Lai, H. T. T., Trung, H. D., Lai, K. L., & Nguyen, D.-T. (2024). Enhancing grid stability through predictive control and fuzzy neural networks in flywheel energy storage systems integration.

International Journal of Modern Physics B. Advance online publication. https://doi.org/10.1142/S021797922540020X

Miranda, R. L., Fernandes, L. D. E. S., Araujo, D. N., Medrado, R. C., Filho, D. F. G., & Junior, A. G. M. (2023, November). Application and validation of WECC models for photovoltaic and energy storage systems in ANATEM software for power system stability studies. In *2023 IEEE PES Innovative Smart Grid Technologies Latin America (ISGT Latin America)* (pp. 210–214). IEEE.

Monaghan, R. F. D., Kumar, M., Singer, S. L., Zhang, C., & Ghoniem, A. F. (2010). Reduced order modeling of entrained flow solid fuel gasification. In *ASME 2009 International Mechanical Engineering Congress and Exposition* (Vol. 3, pp. 397–409). American Society of Mechanical Engineers Digital Collection.

Nezhad, A. E., Nardelli, P. H. J., Sahoo, S., Ghanavati, F., & Osório, G. J. (2022). A centralized home energy management system to minimize consumer's electricity bill. In *2022 IEEE International Conference on Environment and Electrical Engineering and 2022 IEEE Industrial and Commercial Power Systems Europe (EEEIC/I&CPS Europe)* (pp. 1–5). IEEE.

Nguyen, V. T., Nguyen, H. H., Le, K. H., & Truong, T. K. (2021). Expansion of renewable energy capacities in microgrids using robust control approaches. *GMSARN International Journal, 15*(4), 336–344.

Perinov, Hudaya, C., Sudiarto, B., Garniwa, I. (2021). Effects of solar and wind power energy sources integration on frequency dynamics in microgrid. *Journal of Physics: Conference Series, 1858*(1), 012047.

Saritha, K. S., Sreedharan, S., Anand, S. R., & Nair, U. (2021). Swarm intelligence approach for optimum renewable integration of campus microgrid-a case study. *GMSARN International Journal,15*(2), 157–165.

Sattar, F., Ghosh, S., Isbeih, Y. J., El Moursi, M. S., Al Durra, A., & El Fouly, T. H. (2024). A predictive tool for power system operators to ensure frequency stability for power grids with renewable energy integration. *Applied Energy, 353*, 122226.

Svendsen, H. G., Vrana, T. K., Holdyk, A., & Schümann, H. (2023). The low emission oil and gas open reference platform-an off-grid energy system for renewable integration studies. *IET Energy Systems Integration*, *5*(1), 66–79.

Zhangaskanov, D., Sagidolla, B., Bagdaulet, K., Yong, Z., Xiaohui, S., & Eddieyin, K. (2022). High-fidelity 2-way FSI simulation of a wind turbine using fully structured multiblock meshes in OpenFoam for accurate aero-elastic analysis. *Fluids*, *7*(5), 169.

Zhao, F., Li, Y., Wei, Y., Wang, D., Li, F., & Yang, X. (2022, October). Integrated energy production unit capacity optimization and year-round operation simulation. In *2022 IEEE 7th International Conference on Power and Renewable Energy (ICPRE)* (pp. 1206–1210). IEEE.

The manufacturer's authorised representative in the EU is Springer Nature Customer Service Centre GmbH, Europaplatz 3, 69115 Heidelberg, Germany. If you have any concerns regarding our products, please contact ProductSafety@springernature.com

Printed and bound by CPI Group (UK) Ltd, Croydon, CR0 4YY

30/09/2026

02225300-0010